LONDON MATHEMATICAL SOCIETY LECTURE NOTE SERIES

The titles below are available from booksellers, or from Cambridge University Press at www.cambridge.org/mathematics

384 Motivic integration and its interactions with model theory and non-Archimedean geometry I, R. CLUCKERS, J. NICAISE & J. SEBAG (eds)
385 Entropy of hidden Markov processes and connections to dynamical systems, B. MARCUS, K. PETERSEN & T. WEISSMAN (eds)
386 Independence-friendly logic, A.L. MANN, G. SANDU & M. SEVENSTER
387 Groups St Andrews 2009 in Bath I, C.M. CAMPBELL *et al* (eds)
388 Groups St Andrews 2009 in Bath II, C.M. CAMPBELL *et al* (eds)
389 Random fields on the sphere, D. MARINUCCI & G. PECCATI
390 Localization in periodic potentials, D.E. PELINOVSKY
391 Fusion systems in algebra and topology, M. ASCHBACHER, R. KESSAR & B. OLIVER
392 Surveys in combinatorics 2011, R. CHAPMAN (ed)
393 Non-abelian fundamental groups and Iwasawa theory, J. COATES *et al* (eds)
394 Variational problems in differential geometry, R. BIELAWSKI, K. HOUSTON & M. SPEIGHT (eds)
395 How groups grow, A. MANN
396 Arithmetic differential operators over the p-adic integers, C.C. RALPH & S.R. SIMANCA
397 Hyperbolic geometry and applications in quantum chaos and cosmology, J. BOLTE & F. STEINER (eds)
398 Mathematical models in contact mechanics, M. SOFONEA & A. MATEI
399 Circuit double cover of graphs, C.-Q. ZHANG
400 Dense sphere packings: a blueprint for formal proofs, T. HALES
401 A double Hall algebra approach to affine quantum Schur–Weyl theory, B. DENG, J. DU & Q. FU
402 Mathematical aspects of fluid mechanics, J.C. ROBINSON, J.L. RODRIGO & W. SADOWSKI (eds)
403 Foundations of computational mathematics, Budapest 2011, F. CUCKER, T. KRICK, A. PINKUS & A. SZANTO (eds)
404 Operator methods for boundary value problems, S. HASSI, H.S.V. DE SNOO & F.H. SZAFRANIEC (eds)
405 Torsors, étale homotopy and applications to rational points, A.N. SKOROBOGATOV (ed)
406 Appalachian set theory, J. CUMMINGS & E. SCHIMMERLING (eds)
407 The maximal subgroups of the low-dimensional finite classical groups, J.N. BRAY, D.F. HOLT & C.M. RONEY-DOUGAL
408 Complexity science: the Warwick master's course, R. BALL, V. KOLOKOLTSOV & R.S. MACKAY (eds)
409 Surveys in combinatorics 2013, S.R. BLACKBURN, S. GERKE & M. WILDON (eds)
410 Representation theory and harmonic analysis of wreath products of finite groups, T. CECCHERINI-SILBERSTEIN, F. SCARABOTTI & F. TOLLI
411 Moduli spaces, L. BRAMBILA-PAZ, O. GARCÍA-PRADA, P. NEWSTEAD & R.P. THOMAS (eds)
412 Automorphisms and equivalence relations in topological dynamics, D.B. ELLIS & R. ELLIS
413 Optimal transportation, Y. OLLIVIER, H. PAJOT & C. VILLANI (eds)
414 Automorphic forms and Galois representations I, F. DIAMOND, P.L. KASSAEI & M. KIM (eds)
415 Automorphic forms and Galois representations II, F. DIAMOND, P.L. KASSAEI & M. KIM (eds)
416 Reversibility in dynamics and group theory, A.G. O'FARRELL & I. SHORT
417 Recent advances in algebraic geometry, C.D. HACON, M. MUSTAŢĂ & M. POPA (eds)
418 The Bloch–Kato conjecture for the Riemann zeta function, J. COATES, A. RAGHURAM, A. SAIKIA & R. SUJATHA (eds)
419 The Cauchy problem for non-Lipschitz semi-linear parabolic partial differential equations, J.C. MEYER & D.J. NEEDHAM
420 Arithmetic and geometry, L. DIEULEFAIT *et al* (eds)
421 O-minimality and Diophantine geometry, G.O. JONES & A.J. WILKIE (eds)
422 Groups St Andrews 2013, C.M. CAMPBELL *et al* (eds)
423 Inequalities for graph eigenvalues, Z. STANIĆ
424 Surveys in combinatorics 2015, A. CZUMAJ *et al* (eds)
425 Geometry, topology and dynamics in negative curvature, C.S. ARAVINDA, F.T. FARRELL & J.-F. LAFONT (eds)
426 Lectures on the theory of water waves, T. BRIDGES, M. GROVES & D. NICHOLLS (eds)
427 Recent advances in Hodge theory, M. KERR & G. PEARLSTEIN (eds)
428 Geometry in a Fréchet context, C.T.J. DODSON, G. GALANIS & E. VASSILIOU
429 Sheaves and functions modulo p, L. TAELMAN
430 Recent progress in the theory of the Euler and Navier–Stokes equations, J.C. ROBINSON, J.L. RODRIGO, W. SADOWSKI & A. VIDAL-LÓPEZ (eds)
431 Harmonic and subharmonic function theory on the real hyperbolic ball, M. STOLL
432 Topics in graph automorphisms and reconstruction (2nd Edition), J. LAURI & R. SCAPELLATO
433 Regular and irregular holonomic D-modules, M. KASHIWARA & P. SCHAPIRA
434 Analytic semigroups and semilinear initial boundary value problems (2nd Edition), K. TAIRA

435 Graded rings and graded Grothendieck groups, R. HAZRAT
436 Groups, graphs and random walks, T. CECCHERINI-SILBERSTEIN, M. SALVATORI & E. SAVA-HUSS (eds)
437 Dynamics and analytic number theory, D. BADZIAHIN, A. GORODNIK & N. PEYERIMHOFF (eds)
438 Random walks and heat kernels on graphs, M.T. BARLOW
439 Evolution equations, K. AMMARI & S. GERBI (eds)
440 Surveys in combinatorics 2017, A. CLAESSON *et al* (eds)
441 Polynomials and the mod 2 Steenrod algebra I, G. WALKER & R.M.W. WOOD
442 Polynomials and the mod 2 Steenrod algebra II, G. WALKER & R.M.W. WOOD
443 Asymptotic analysis in general relativity, T. DAUDÉ, D. HÄFNER & J.-P. NICOLAS (eds)
444 Geometric and cohomological group theory, P.H. KROPHOLLER, I.J. LEARY, C. MARTÍNEZ-PÉREZ & B.E.A. NUCINKIS (eds)
445 Introduction to hidden semi-Markov models, J. VAN DER HOEK & R.J. ELLIOTT
446 Advances in two-dimensional homotopy and combinatorial group theory, W. METZLER & S. ROSEBROCK (eds)
447 New directions in locally compact groups, P.-E. CAPRACE & N. MONOD (eds)
448 Synthetic differential topology, M.C. BUNGE, F. GAGO & A.M. SAN LUIS
449 Permutation groups and cartesian decompositions, C.E. PRAEGER & C. SCHNEIDER
450 Partial differential equations arising from physics and geometry, M. BEN AYED *et al* (eds)
451 Topological methods in group theory, N. BROADDUS, M. DAVIS, J.-F. LAFONT & I. ORTIZ (eds)
452 Partial differential equations in fluid mechanics, C.L. FEFFERMAN, J.C. ROBINSON & J.L. RODRIGO (eds)
453 Stochastic stability of differential equations in abstract spaces, K. LIU
454 Beyond hyperbolicity, M. HAGEN, R. WEBB & H. WILTON (eds)
455 Groups St Andrews 2017 in Birmingham, C.M. CAMPBELL *et al* (eds)
456 Surveys in combinatorics 2019, A. LO, R. MYCROFT, G. PERARNAU & A. TREGLOWN (eds)
457 Shimura varieties, T. HAINES & M. HARRIS (eds)
458 Integrable systems and algebraic geometry I, R. DONAGI & T. SHASKA (eds)
459 Integrable systems and algebraic geometry II, R. DONAGI & T. SHASKA (eds)
460 Wigner-type theorems for Hilbert Grassmannians, M. PANKOV
461 Analysis and geometry on graphs and manifolds, M. KELLER, D. LENZ & R.K. WOJCIECHOWSKI
462 Zeta and L-functions of varieties and motives, B. KAHN
463 Differential geometry in the large, O. DEARRICOTT *et al* (eds)
464 Lectures on orthogonal polynomials and special functions, H.S. COHL & M.E.H. ISMAIL (eds)
465 Constrained Willmore surfaces, Á.C. QUINTINO
466 Invariance of modules under automorphisms of their envelopes and covers, A.K. SRIVASTAVA, A. TUGANBAEV & P.A. GUIL ASENSIO
467 The genesis of the Langlands program, J. MUELLER & F. SHAHIDI
468 (Co)end calculus, F. LOREGIAN
469 Computational cryptography, J.W. BOS & M. STAM (eds)
470 Surveys in combinatorics 2021, K.K. DABROWSKI *et al* (eds)
471 Matrix analysis and entrywise positivity preservers, A. KHARE
472 Facets of algebraic geometry I, P. ALUFFI *et al* (eds)
473 Facets of algebraic geometry II, P. ALUFFI *et al* (eds)
474 Equivariant topology and derived algebra, S. BALCHIN, D. BARNES, M. KEDZIOREK & M. SZYMIK (eds)
475 Effective results and methods for Diophantine equations over finitely generated domains, J.-H. EVERTSE & K. GYORY
476 An indefinite excursion in operator theory, A. GHEONDEA
477 Elliptic regularity theory by approximation methods, E.A. PIMENTEL
478 Recent developments in algebraic geometry, H. ABBAN, G. BROWN, A. KASPRZYK & S. MORI (eds)
479 Bounded cohomology and simplicial volume, C. CAMPAGNOLO, F. FOURNIER- FACIO, N. HEUER & M. MORASCHINI (eds)
480 Stacks Project Expository Collection (SPEC), P. BELMANS, W. HO & A.J. DE JONG (eds)
481 Surveys in combinatorics 2022, A. NIXON & S. PRENDIVILLE (eds)
482 The logical approach to automatic sequences, J. SHALLIT
483 Rectifiability: a survey, P. MATTILA
484 Discrete quantum walks on graphs and digraphs, C. GODSIL & H. ZHAN
485 The Calabi problem for Fano threefolds, C. ARAUJO *et al*
486 Modern trends in algebra and representation theory, D. JORDAN, N. MAZZA & S. SCHROLL (eds)
487 Algebraic combinatorics and the Monster group, A.A. IVANOV (ed)
488 Maurer–Cartan methods in deformation theory, V. DOTSENKO, S. SHADRIN & B. VALLETTE
489 Higher dimensional algebraic geometry, C. HACON & C. XU (eds)
490 C^∞-algebraic geometry with corners, K. FRANCIS-STAITE & D. JOYCE
491 Groups and graphs, designs and dynamics, R.A. BAILEY, P.J. CAMERON & Y. WU (eds)
492 Homotopy theory of enriched Mackey functors, N. JOHNSON & D. YAU
493 Surveys in combinatorics 2024, F. FISCHER & R. JOHNSON (eds)
494 K-theory and representation theory, R. PLYMEN & M.H. ŞENGÜN (eds)

London Mathematical Society Lecture Note Series: 494

K-Theory and Representation Theory

Edited by

ROGER PLYMEN
University of Manchester

MEHMET HALUK ŞENGÜN
University of Sheffield

Shaftesbury Road, Cambridge CB2 8EA, United Kingdom

One Liberty Plaza, 20th Floor, New York, NY 10006, USA

477 Williamstown Road, Port Melbourne, VIC 3207, Australia

314–321, 3rd Floor, Plot 3, Splendor Forum, Jasola District Centre,
New Delhi – 110025, India

103 Penang Road, #05–06/07, Visioncrest Commercial, Singapore 238467

Cambridge University Press is part of Cambridge University Press & Assessment,
a department of the University of Cambridge.

We share the University's mission to contribute to society through the pursuit of
education, learning and research at the highest international levels of excellence.

www.cambridge.org
Information on this title: www.cambridge.org/9781009201506

DOI: 10.1017/9781009201476

When citing this work, please include a reference to the DOI 10.1017/9781009201476

First published 2025

Printed in the United Kingdom by TJ Books Limited, Padstow Cornwall

A catalogue record for this publication is available from the British Library

Library of Congress Cataloging-in-Publication Data
Names: Plymen, Roger J., editor. | Şengün, Mehmet Haluk, editor.
Title: K-theory and representation theory / edited by Roger Plymen,
University of Manchester, Mehmet Haluk Şengün, University of Sheffield.
Description: Cambridge, United Kingdom ; New York, NY : Cambridge
University Press, 2024. | Series: London Mathematical Society lecture
note series ; 494 | Includes bibliographical references and index.
Identifiers: LCCN 2024012153 | ISBN 9781009201506 (paperback) |
ISBN 9781009201476 (ebook)
Subjects: LCSH: K-theory. | Representations of groups. |
Representations of Lie groups.
Classification: LCC QA612.33 .K28 2024 | DDC 512/.66–dc23/eng20240723
LC record available at https://lccn.loc.gov/2024012153

ISBN 978-1-009-20150-6 Paperback

Contents

Contributors

Anne-Marie Aubert
Sorbonne Université and Université Paris Cité, CNRS, IMJ-PRG

Peter Hochs
Radboud University

Bram Mesland
Leiden University

Mehmet Haluk Şengün
University of Sheffield

Hang Wang
East China Normal University

Preface

In July 2021, three of us, Nigel Higson, Roger Plymen, and Haluk Şengün, organized an online summer school on K-theory and representation theory. The original intention was for this to be an in-person event. However, the constraints of the pandemic forced us to have the event online.

The lecturers, Anne-Marie Aubert, Peter Hochs, Bram Mesland, and Hang Wang, gave their minicourses comprising four lectures each. Much time has elapsed, and the lecturers have taken the opportunity to considerably expand their lecture notes into their present forms. It is our sincere hope that these lecture notes will prove useful to beginning graduate students who wish to enter this area, particularly because much of the material collated here is scattered around the literature.

In Chapter 1, Mesland and Şengün set the scene by describing the basics of the theory of C^*-algebras associated with locally compact groups. The important notions of Hilbert C^*-modules, C^*-correspondences, and their associated generalized induction functors are discussed. After the introduction of Morita equivalence, examples of C^*-correspondences that capture well-known constructions from representation theory are given.

The authors then move onto a detailed discussion of K-theory and its powerful generalization KK-theory, finishing with a basic discussion of Dirac induction. This is hopefully a good foundation for Chapter 3. A first course in functional analysis is a useful prerequisite.

In Chapter 2, Hochs describes the classification of tempered representations of semisimple Lie groups. The treatment starts from the very basic notions, supporting the reader with concrete examples throughout the journey. The structure theorem of the reduced group C^*-algebra of a real linear reductive group and its K-theory is described based on the recent accounts of Nigel Higson and collaborators.

In Chapter 3, Wang describes with many examples the role of the Dirac operator in the representation theory of semisimple Lie groups. This topic has a long history with roots going back to holomorphic induction. She develops the subject carefully with links to the previous chapters. A first course in differential geometry would be a useful prerequisite.

In Chapter 4, Aubert presents an essentially self-contained introduction to the representation theory of p-adic groups. After a formulation of the Aubert–Baum–Plymen–Solleveld conjecture, her chapter culminates in a discussion of her recent work with Afgoustidis, which provides, under a suitable hypothesis, a structure theorem for the reduced group C^*-algebra of a (linear) reductive group over any local field of characteristic zero.

1

Group C^*-Algebras, C^*-Correspondences and K-Theory

Bram Mesland and Mehmet Haluk Şengün

1.1 Introduction

It is an elementary fact that representations of a finite group can be viewed as modules over its group algebra and vice versa. If we consider the unitary representations of a locally compact Hausdorff group G, the role of the group algebra is played by a certain C^*-algebra (an involutive Banach algebra whose norm satisfies the so-called C^*-identity) called the (maximal) group C^*-algebra of G denoted as $C^*(G)$. Introduced in the late 1940s, $C^*(G)$ is obtained by taking the completion of the convolution algebra $L^1(G)$ with respect to a norm that takes into account all of the unitary dual of G. Very soon after their introductions, the C^*-algebra $C^*(G)$ and its quotient $C^*_r(G)$ (the reduced group C^*-algebra) were used to prove the existence of direct integral decompositions of unitary representations and of the existence of the abstract Plancherel measure for Type I groups (see e.g. [7, Section 18] or [9, Section 7]).

In the 1970s, Rieffel introduced the notion of C^*-correspondence for C^*-algebras and the associated generalized induction functor on their categories of representations. Rieffel also identified the conditions required on these C^*-correspondences for the associated functors to be invertible, thus obtaining a C^*-algebraic version of the notion of Morita equivalence from ring theory. Specialized to the setting of group C^*-algebras, Rieffel's work elegantly captures Mackey's theory of induction, in particular, the imprimitivity theorem, for unitary representations of locally compact Hausdorff groups (which were arguably the main motivations for Rieffel). Since their introduction, these notions have become fundamental tools in the study of C^*-algebras.

The study of K-theory for C^*-algebras concerns invariants defined up to a suitable notion of homotopy. The fact that Morita equivalent C^*-algebras have isomorphic K-theory adds weight to the viewpoint of Morita equivalence as the correct categorical notion of isomorphism for C^*-algebras. In representation theory, studying the K-theory of $C^*(G)$ can be viewed as studying the homology of the typically non-Hausdorff unitary dual of a locally compact group G.

In these notes, we aim to discuss the basics of group C^*-algebras touching upon some of the themes mentioned earlier, and with the goal of providing (most of) the background that is required to understand the C^*-algebraic notions that appear in the other lectures of these proceedings. As such, after introducing group C^*-algebras, we move on to define the crossed product C^*-algebra construction that captures group actions on topological spaces. We then discuss C^*-correspondences and their associated generalized induction functors. In particular, we discuss how one captures Mackey's induction functor. We finish that section with a short discussion of a more recent result concerning the local theta correspondence of Roger Howe as a generalized induction functor associated with a C^*-correspondence of group C^*-algebras. Our final topic is K-theory. We give an overview of the main structural properties of K-theory, as well as a short discussion of Kasparov's equivariant KK-theory and the Connes–Kasparov conjecture.

1.2 C^*-Algebras and Locally Compact Groups

1.2.1 C^*-Algebras

The basic objects of noncommutative geometry are C^*-algebras, a special class of associative Banach algebras over the field of complex numbers, which enjoys rigid properties. Recall that a *Banach algebra* is an associative algebra A equipped with a norm in which it is a Banach space such that the norm and the multiplication satisfy $\|ab\| \leq \|a\|\|b\|$. Furthermore, an *involution* on a complex associative algebra is a conjugate linear map $a \mapsto a^*$ such that $(ab)^* = b^*a^*$ and $(a^*)^* = a$. A $*$-*homomorphism* between Banach $*$-algebras A and B is an algebra homomorphism $\phi : A \to B$ for which $\phi(a^*) = \phi(a)^*$. Among Banach algebras with involution, the class of C^*-algebras is characterized by imposing an additional relation between the norm and the involution.

Definition 1.2.2. A C^*-algebra is a Banach algebra A with involution $a \mapsto a^*$ such that the involution satisfies the C^*-*identity*: $\|a^*a\| = \|a\|^2$.

One easily checks that for a Hilbert space H, any norm-closed $*$-subalgebra of $\mathbb{B}(H)$ is a C^*-algebra. Such C^*-algebras are sometimes referred to as *concrete* C^*-algebras. Also, it must be noted that $*$-homomorphisms are automatically norm contractive, and injective $*$-homomorphisms are isometric.

One of the cornerstones of the theory of C^*-algebras is the Gelfand–Naimark–Segal (GNS) theorem. This theorem states that, in fact, any C^*-algebra is a concrete C^*-algebra:

Theorem 1.2.3 (GNS). *Let A be a C^*-algebra. There exists a Hilbert space H and an injective $*$-homomorphism $\phi : A \to \mathbb{B}(H)$.*

This theorem is mainly useful for theoretical purposes. The construction of the Hilbert space H is canonical, but in examples, it is often not the most efficient choice.

The starting point for the field of noncommutative geometry is the Gelfand–Naimark theorem describing the duality between compact Hausdorff spaces and commutative C^*-algebras. This duality is implemented by associating a space X with its function algebra:

$$X \mapsto C(X) := \{f : X \to \mathbb{C} : f \quad \text{is continuous}\},$$

which is a C^*-algebra in the sup-norm with pointwise multiplication of functions and involution given by pointwise complex conjugation. The inverse map is given by taking the space of maximal ideals or characters of a commutative C^*-algebra:

$$A \mapsto \mathrm{Max}_A := \{I \subset A : I \quad \text{maximal ideal}\} \simeq \{\phi : A \to \mathbb{C} : \phi(ab) = \phi(a)\phi(b)\},$$

where we note that maximal ideals are automatically norm and *-closed, or equivalently, a character is automatically continuous and a *-homomorphism. The topology on Max_A is the weak *-topology on the space of characters, that is,

$$\phi_n \to \phi \Leftrightarrow \forall a \in A : \phi_n(a) \to \phi(a).$$

Theorem 1.2.4 (Gelfand–Naimark). *Let A be a commutative C^*-algebra with unit. Then A is $*$-isomorphic to $C(\mathrm{Max}_A)$.*

This theorem characterizes commutative unital C^*-algebras completely.

1.2.5 The Dual of a C^*-Algebra

A *representation* of a C^*-algebra is a $*$-homomorphism $\pi : A \to B(H_\pi)$, with H_π being a Hilbert space. Such a representation is *irreducible* if the only $\pi(A)$-invariant subspaces of H_π are 0 and H_π itself. Two representations (π, H_π) and (ρ, H_ρ) are *equivalent* if there exists a unitary isomorphism $H_\pi \to H_\rho$ intertwining the A representations π and ρ. In view of the GNS theorem, C^*-algebras have a rich representation theory. This allows us to associate various topological spaces with a possibly noncommutative C^*-algebra. The *spectrum* of A is the set of equivalence classes of irreducible representations of A:

$$\widehat{A} := \{[(\pi, V_\pi)] : \pi : A \to \mathbb{B}(V_\pi)\}.$$

Definition 1.2.6. Let A be a C^*-algebra. The *ideal lattice* $\mathscr{I}(A)$ is the set of all closed two-sided ideals in A. The *primitive ideal spectrum* is defined as the space of ideals

$$\text{Prim } A := \left\{I \subset A : \exists \pi \in \widehat{A},\ I = \ker \pi\right\} = \left\{\ker \pi : \pi \in \widehat{A}\right\},$$

which is a subset of $\mathscr{I}(A)$.

The ideal lattice $\mathscr{I}(A)$ is a topological space when equipped with the hull-kernel or Jacobson topology: The closure of a collection $\{J_\lambda\}_{\lambda \in \Lambda}$ is defined to be

$$\overline{\{J_\lambda\}}_{\lambda \in \Lambda} := \left\{J : J \supset \bigcap_{\lambda \in \Lambda} J_\lambda\right\}.$$

The primitive ideal spectrum Prim $A \subset \mathscr{I}(A)$ inherits the relative topology.

For commutative C^*-algebras, we have Max A = Prim A and thus $A \simeq C_0(\text{Prim } A)$, but in general Prim A is only a T_0-space.

There is a natural map

$$\widehat{A} \to \text{Prim } A, \quad [(\pi, V_\pi)] \mapsto \ker \pi,$$

and the space $\widehat{A}$ is topologized by declaring this map to be continuous.

1.2.7 Group C^*-Algebras and Unitary Representations

Let G be a locally compact Hausdorff topological group. With such a group, one can associate various operator algebras that encode the representation theory of G. For simplicity, we will furthermore assume that G is unimodular and let μ denote the bi-invariant Haar measure on G. The Banach space $L^1(G) := L^1(G, \mu)$ is a Banach $*$-algebra for the convolution product and involution

$$f * g(t) := \int_G f(s) g(s^{-1}t) \mathrm{d}s, \quad f^*(t) = \overline{f(t^{-1})}.$$

Denote by $\widehat{G}$ the set of isomorphism classes of irreducible unitary representations of G and let $\pi : G \to \mathbb{U}(H_\pi)$ be such a representation. Its *integrated form* is the map

$$\pi : L^1(G) \to \mathbb{B}(H_\pi), \quad \pi(f) \cdot v := \int_G f(s) \pi(s) \cdot v \mathrm{d}s,$$

which is a contractive $*$-homomorphism. The following result is well known.

Proposition 1.2.8. *Let G be a locally compact (unimodular) group and H a Hilbert space. There is a 1–1 correspondence between unitary representations $\pi : G \to \mathbb{U}(H)$ and contractive $*$-homomorphisms $\pi : L^1(G) \to \mathbb{B}(H)$.*

As illustrated by Proposition 1.2.8, the Hilbert space representation theory of $L^1(G)$ is intimately linked to that of G. However, $L^1(G)$ is not a C^*-algebra in general: its norm and involution do not satisfy the C^*-identity. We can nonetheless construct C^*-algebras from $L^1(G)$.

Definition 1.2.9. The *full C^*-norm* on $L^1(G)$ is defined by

$$\|f\|_{C^*(G)} := \sup\left\{\|\pi(f)\|_{\mathbb{B}(H_\pi)} : \pi \in \widehat{G}\right\}.$$

Note that $\|f\|_{C^*(G)}$ is a C^*-norm and $\|f\|_{C^*(G)} \leq \|f\|_{L^1(G)}$. The *full group C^*-algebra* $C^*(G)$ is defined to be the completion of $L^1(G)$ in the C^*-norm $\|\cdot\|_{C^*(G)}$.

By construction, a unitary representation $\pi : G \to \mathbb{U}(H_\pi)$ gives rise to a $*$-representation $\pi : C^*(G) \to \mathbb{B}(H_\pi)$, and π is irreducible as a G-representation if and only if it is irreducible as a $C^*(G)$ representation. The following result is classical; see, for example, [4, Section 8.B].

Theorem 1.2.10. *The natural map that takes a unitary representation of G to its integrated form on $C^*(G)$ establishes a homeomorphism between $\widehat{G}$ and $\widehat{C^*(G)}$ with respect to their natural topologies.*

Given π, we denote the image of $C^*(G)$ inside $\mathbb{B}(H_\pi)$ by $C^*_\pi(G)$.

Proposition 1.2.11. *The dual $\widehat{C^*_\pi(G)}$ is homeomorphic to the support of π.*

Recall that the *support* of a unitary representation σ of G is the set of irreducible unitary representations of G that are weakly contained in σ.

Of particular interest is the *left regular representation* of G on $L^2(G)$.

Definition 1.2.12. The *reduced C^*-algebra* of G is the image of $C^*(G)$ inside $\mathbb{B}(L^2(G))$ and is denoted as $C^*_r(G)$.

It follows from Proposition 1.2.11 that the spectrum $\widehat{C^*_r(G)}$ is the support of the regular representation of G. In the context of representation theory of semisimple groups, representations weakly contained in the regular representation are called *tempered*. In particular, the reduced group C^*-algebra of a semisimple algebraic group G over a local field captures the tempered representations of G.

1.2.13 Group Actions on C^*-Algebras

For a C^*-algebra A, write

$$\mathrm{Aut}\, A := \{\alpha : A \to A : \alpha \text{ is a } *\text{-automorphism}\}$$

for its automorphism group.

Definition 1.2.14. Let G be a locally compact group and A be a C^*-algebra. A *strongly continuous action of* G *on* A is a group homomorphism

$$G \to \mathrm{Aut}\, A, \quad g \mapsto \alpha_g,$$

such that for all $a \in A$, the map

$$G \to A, \quad g \mapsto \alpha_g(a)$$

is continuous for the norm topology on A. We refer to (A, α) as a *G-C^*-algebra.*

Example 1.2.15. Let X be a locally compact Hausdorff space and suppose that G acts on X by homeomorphisms in such a way that the map

$$G \times X \to X \times X, \quad (g, x) \mapsto (x, g \cdot x)$$

is continuous. Then the induced action of G on $C_0(X)$,

$$g \cdot f(x) := f(g^{-1} \cdot x), \quad g \in G, \ f \in C_0(X)$$

is a strongly continuous action of G on $C_0(X)$.

Given a strongly continuous action of G on A, the subspace

$$A^G := \left\{a \in A : \forall g \in G \quad \alpha_g(a) = a\right\}$$

is a C^*-subalgebra of A called the *fixed-point algebra.*

Definition 1.2.16. Let A be a C^*-algebra and $B \subset A$ a C^*-subalgebra. A *conditional expectation of* B *onto* A is a map $\rho : A \to B$ satisfying

$$\rho(a^* a) \geq 0, \quad \rho(ab) = \rho(a)b,$$

for all $a \in A$ and $b \in B$.

When G is compact and μ is the normalized Haar measure on G, the map

$$a \mapsto \int_G \alpha_g(a) \mathrm{d}\mu(g)$$

is a conditional expectation from A onto A^G. In case A is unital, $\mathbb{C} \cdot 1 \subset A^G$. In general, when G is noncompact and A is nonunital, we may have that $A^G = 0$. This holds for instance for the left-translation action of a noncompact group G on itself: a translation invariant function $f \in C_0(G)$ which vanishes at infinity has to be the zero function.

1.3 Hilbert C^*-Modules

Standard references for the materials in this section are [17, 20, 23].

1.3.1 Hilbert C^*-Modules

A noncommutative algebra may have very few or no two-sided ideals at all. The algebras $M_n(\mathbb{C})$ of $n \times n$-matrices, and the algebra $\mathbb{K}$ of compact operators on a (separable) Hilbert space, are examples of such *simple* algebras.

The kernel of a homomorphism $\phi : A \to B$ is a two-sided ideal. Hence if A is a simple algebra, the map ϕ is either zero or injective. This indicates that the notion of homomorphism is not rich enough for noncommutative algebras.

The homomorphism ϕ can also be interpreted as equipping the right B-module B with a left A-module structure. It turns out to be fruitful to view bimodules as morphisms between algebras.

Definition 1.3.2. Let B be a C^*-algebra with unit. A finitely generated B-module X is *projective*, if there exists a right B-module Y and an $n \in \mathbb{N}$, such that $X \oplus Y \cong B^n$ as right B-modules.

This definition is motivated by the following classical result.

Theorem 1.3.3 (Serre–Swan). *Let M be a compact Hausdorff space and $\mathscr{E} \xrightarrow{\pi} M$ a locally trivial finite-dimensional complex vector bundle. The module of continuous sections*

$$\phi : M \to \mathscr{E}, \quad \pi \circ \phi = \mathrm{id}_X,$$

is a finitely generated projective $C(M)$-module. Conversely, every locally trivial finite-dimensional vector bundle over M arises in this way.

This theorem is proved by showing that $\mathscr{E}$ is a direct summand in a trivial bundle $M \times \mathbb{C}^n$.

Definition 1.3.4. Let B be a C^*-algebra and X a right B-module. Then X is a (right) *Hilbert C^*-module* over B provided that there is a right $\mathbb{C}$-linear pairing

$$X \times X \to B$$
$$(x_1, x_2) \mapsto \langle x_1, x_2 \rangle,$$

with the following properties:

1. $\langle x_1, x_2 b\rangle = \langle x_1, x_2\rangle b$ for all $b \in B$;
2. $\langle x_1, x_2\rangle^* = \langle x_2, x_1\rangle$ for all $x_i \in X$;
3. $\langle x, x\rangle \geq 0$ in B for all $x \in X$;
4. X is complete with respect to the norm $\|x\|^2 := \|\langle x, x\rangle\|$.

The module X is called *full* if the span of elements $\langle x_1, x_2 \rangle$ is dense in B. The notion of a left Hilbert C^*-module is defined similarly.

Example 1.3.5. Let $\mathscr{E} \to M$ be a vector bundle. A Riemannian metric is nothing but a $C(M)$-valued inner product $\Gamma(\mathscr{E}) \times \Gamma(\mathscr{E}) \to C(M)$. Any choice of Riemannian metric on $\mathscr{E}$ makes $\Gamma(\mathscr{E})$ into a Hilbert C^*-module over $C(M)$.

Example 1.3.6. For all n, the module B^n admits the inner product

$$\langle (a_i), (b_i) \rangle := \sum a_i^* b_i,$$

and it is not hard to check that B is a C^*-module for this inner product. Consequently, every finitely generated projective module X admits an inner product, using an embedding $X \to B^n$.

The converse is also true.

Proposition 1.3.7. *Let X be a finitely generated full Hilbert C^*-module over a unital C^*-algebra B. Then X is projective.*

A direct proof of this fact can be found in [10, Theorem 5.9]. A Hilbert C^*-module X is *countably generated* if there is a countable subset of X that generates a dense submodule. The analogue of Proposition 1.3.7 for countably generated Hilbert C^*-modules requires an analytic version of the notion of free module.

Definition 1.3.8. Let A be a C^*-algebra. The *standard module* over A is the space

$$H_A := \left\{ (a_i)_{i=1}^{\infty} : \sum a_i^* a_i < \infty \right\},$$

with convergence of the series in A.

The standard module H_A is a Hilbert C^*-module in the inner product

$$\langle (a_i), (b_i) \rangle := \sum a_i^* b_i.$$

There is a natural notion of direct sum for Hilbert C^*-modules X and Y over a C^*-algebra B. On the B-module $X \oplus Y$, the pairing

$$\langle (x_1, y_1), (x_2, y_2) \rangle := \langle x_1, x_2 \rangle + \langle y_1, y_2 \rangle$$

equips $X \oplus Y$ with the structure of a Hilbert C^*-module. An isomorphism of Hilbert C^*-modules is a module isomorphism that intertwines the inner products. Countably generated Hilbert C^*-modules are projective in the following sense.

Theorem 1.3.9 (Kasparov stabilization theorem [14]). *Let X be a countably generated Hilbert C^*-module over a σ-unital C^*-algebra A. There is an isomorphism $X \oplus H_A \xrightarrow{\sim} H_A$.*

We note that the isomorphism in this theorem is not unique. Also, historically, the first proofs of Proposition 1.3.7 used Theorem 1.3.9 in combination with [20, Theorem 2.7.5].

1.3.10 Operators on Hilbert C^*-Modules

Given a Hilbert C^*-module X over B, we define

$$\mathrm{End}_B^*(X) := \{T : X \to X : \exists\ T^* : X \to X, \quad \langle Tx_1, x_2\rangle = \langle x_1, T^*x_2\rangle\}.$$

Operators in $\mathrm{End}_B^*(X)$ are B-linear and bounded, and the adjoint T^* is unique and conjugate linear. The $*$-algebra $\mathrm{End}_B^*(X)$ is a C^*-algebra in the operator norm. Similarly, for a pair X, Y of Hilbert C^*-modules over B, one defines the space $\mathrm{Hom}_B^*(X,Y)$. It is useful to observe that $\mathrm{Hom}_B^*(X,Y) \subset \mathrm{End}_B^*(X \oplus Y)$.

The existence of the adjoint is nontrivial: there exist bounded, B-linear maps that are not adjointable. To see this, consider the C^*-algebra $A := C([0,1])$ and its closed two-sided ideal $J := C_0((0,1])$ of functions vanishing at 0. The direct sum

$$X := A \oplus J \subset A \oplus A$$

is a closed submodule of $A \oplus A$ and hence a Hilbert C^*-module over $B := A \oplus A$. Consider the B-linear map

$$T : X \to X, \quad (a, j) \mapsto (j, 0).$$

Suppose T^* is an adjoint for T, and write $T^*(1,0) = (f, g)$. Then for any $j \in J$, we would have

$$\langle T(0,j), (1,0)\rangle = j^* = \langle (0,j), T^*(1,0)\rangle = j^*g.$$

Since j was arbitrary, it follows that g must be the constant function 1, but this function is not in J. So T^* cannot exist.

This phenomenon is closely related to the notion of *complemented submodule*. A closed submodule $X \subset E$ of a C^*-module is *complemented* if the module

$$X^\perp := \{e \in E : \langle x, e\rangle = 0, \quad \text{for all } x \in X\},$$

has the property that any $e \in E$ can be written uniquely as $e = x + x'$ with $x \in X, x' \in X^\perp$. In other words, $E \cong X \oplus X^\perp$ is, an orthogonal direct sum. To see that this, is a nontrivial property, consider again the ideal $J = C_0((0,1])$.

This is a closed submodule of A, but $J^{\perp} = 0$: an element $g \in J^{\perp}$ satisfies $j^*g = 0$ for all $j \in J$. But this means that $j^*(t)g(t) = 0$ for all $t \in (0,1]$, and since j was arbitrary, this implies that $g = 0$ is the zero function. Hence, obviously, $J \oplus J^{\perp} \neq A$.

1.3.11 C^*-Correspondences

Let A and B be two C^*-algebras. As noncommutative rings may admit few homomorphisms between them, it is natural to consider bimodules instead. We introduce a class of bimodules for C^*-algebras that incorporate their additional analytic structure as well.

Definition 1.3.12. A *C^*-correspondence* (X,α) for (A,B) is a right Hilbert C^*-module X over B, which is equipped with a $*$-homomorphism $\alpha : A \to \mathrm{End}^*_B(X)$.

When convenient, we will denote (X,α) simply by X and will call it an *(A,B)-correspondence* for short. The following construction is fundamental and serves as the bimodule analogue of the composition of homomorphisms.

Definition 1.3.13. Let (Y,α) be a (A,B)-correspondence and X a right Hilbert C^*-module over A. The *C^*-module tensor product* $X \otimes_A Y$ is the completion of the algebraic-balanced tensor product $X \otimes^{\mathrm{alg}}_A Y$ in the norm coming from the C-valued inner product defined by

$$\langle x_1 \otimes y_1, x_2 \otimes y_2 \rangle := \langle y_1, \alpha(\langle x_1, x_2 \rangle) y_2 \rangle. \tag{1.3.13.1}$$

The inner product (1.3.13.1) turns out to be nondegenerate, as the *null space*

$$N = \left\{ \xi \in X \otimes^{\mathrm{alg}} Y : \langle \xi, \xi \rangle = 0 \right\}$$

coincides with the *balancing ideal*

$$I = \mathrm{span} \left\{ xa \otimes y - x \otimes \alpha(a)(y) \in X \otimes^{\mathrm{alg}} Y : x \in X, y \in Y, a \in A \right\}.$$

1.3.14 Compact Operators and Morita Equivalence

For a right Hilbert C^*-module X over a C^*-algebra B, there is an important subalgebra of $\mathrm{End}^*_B(X)$, the algebra $\mathbb{K}(X)$ of *compact operators*. It is the closure of the algebra generated by the rank-one operators

$$T_{x_1,x_2} : x \mapsto x_1 \langle x_2, x \rangle.$$

Note that in case $X = H$ is a Hilbert space, this coincides with the usual definition of $\mathbb{K}(H)$. In general, operators in $\mathbb{K}(X)$ are *not* compact in the sense of Banach spaces but should be viewed as a more algebraic generalization of the notion. In case B is unital and X is finitely generated over B, we have $\mathbb{K}_B(X) = \mathrm{End}^*_B(X)$. For the modules B^n, we have $\mathbb{K}(B^n) = M_n(B)$.

Definition 1.3.15. Let A and B be C^*-algebras. Then A and B are *strongly Morita equivalent* if there exists an (A,B)-bimodule ${}_AX_B$, which is a left Hilbert C^*-module over A with full inner product ${}_A\langle\cdot,\cdot\rangle$, a right C^*-module over B with full inner product $\langle\cdot,\cdot\rangle_B$ such that for all $x,y,z \in X$, we have

$$ {}_A\langle x,y\rangle \cdot z = x \cdot \langle y,z\rangle_B. $$

An equivalent characterization of Morita equivalence is that there exists a full right Hilbert C^*-module X_B over B such that A is isomorphic to $\mathbb{K}(X)$.

The equivalence of both definitions follows by observing that for a Hilbert C^*-module X, there is always a full $\mathbb{K}(X)$-valued left inner product $(x_1,x_2) \mapsto T_{x_1,x_2}$. It is not hard to show that with this inner product, X becomes a left-C^*-module over $\mathbb{K}(X)$. If X is full as a module over B, and the compacts with respect to the $\mathbb{K}(X)$ valued inner product are exactly B.

The symmetry of the definition is obtained by considering the *dual module*

$$ X^* := \mathbb{K}(X,B) $$

of compact operators from X to B. Via the inner product pairing and the involution on B, this becomes a left B-module, and hence a right $K_B(X)$ module. The transitivity of the definition is proved by using the C^*-module tensor product.

From the isomorphism $\mathbb{K}_B(X) \simeq X \otimes_B X^*$, it follows that for an (A,B)-Morita equivalence bimodule X, we have isomorphisms

$$ X \otimes_B X^* \simeq A, X^* \otimes_A X \simeq B. $$

Moreover,

$$ \mathbb{K}(X \otimes_B Y) \simeq X \otimes_B \mathbb{K}(Y) \otimes_B X^*, $$

and thus from this it follows that if Y is a (B,C)-Morita equivalence bimodule, then $X \otimes_B Y$ is an (A,C)-Morita equivalence bimodule:

$$ \mathbb{K}(X \otimes_B Y) \simeq X \otimes_B \mathbb{K}(Y) \otimes_B X^* \simeq X \otimes_B B \otimes_B X^* \simeq \mathbb{K}(X) \simeq A. $$

Last, reflexivity of Morita equivalence is obvious, as $\mathbb{K}(A) = A$.

If both A and B have a unit, then they are Morita equivalent if and only if there exists a finite projective B-module X such that $A \simeq \mathbb{K}_B(X) = \mathrm{End}^*_B(X)$. Thus, in this case, A and B are Morita equivalent if and only if there exists a projection $p \in M_n(B)$ such that $A \simeq pM_n(B)p$.

The Morita equivalence of C^*-algebras induces an equivalence of their representation categories (see [23]).

1.3.16 Crossed Products

We now return to G-C^*-algebras. Just as in the Hilbert space case, we say that a unitary representation of G on a Hilbert C^*-module X if *strongly continuous* if for all $x \in X$ the map

$$G \to X, \quad g \mapsto \alpha_g(x)$$

is continuous for the norm topology on X.

Definition 1.3.17. Let (A, α) be a G-C^*-algebra and B a C^*-algebra. A *covariant representation* of (A, α) is a triple (π, u, X) with X a right Hilbert C^*-module over B, $\pi : A \to \mathrm{End}_B^*(X)$ a $*$-homomorphism and $u : G \to \mathbb{U}(X)$ a strongly continuous unitary representation such that

$$u_g \pi(a) u_g^* = \pi(\alpha_g(a))$$

for all $g \in G$ and $a \in A$.

Given a G-algebra A, we consider the space $C_c(G, A)$ of compactly supported A-valued functions on G. This space is a $*$-algebra in the convolution product and involution given by

$$f_1 * f_2(t) := \int_G f_1(s)\alpha_s(f_2(s^{-1}t)), \ \mathrm{d}\mu(s), \quad f^*(s) := \alpha_s(f(s))^*.$$

These formulae extend to the space $L^1(G, A)$.

Proposition 1.3.18. *Let (A, α) be a G-C^*-algebra and (π, u, X) a covariant representation on a Hilbert C^*-module X over B. Then*

$$\pi \rtimes u : C_c(G, A) \to \mathrm{End}_B^*(X)$$

$$\pi \rtimes u(f)(x) := \int_G \pi(f(s))u_s(x)\mathrm{d}s$$

defines a $*$-representation of $C_c(G, A)$ on X.

Using covariant representations, we can equip $C_c(G, A)$ with a C^*-norm and complete it into a C^*-algebra.

Definition 1.3.19. The *full C^*-norm* on $C_c(G, A)$ to be

$$\|f\|_{\mathrm{full}} := \sup \{\|\pi \rtimes u(f)\| : (\pi, u, X) \text{ a covariant representation of } (A, \alpha)\}.$$

The completion of $C_c(G, A)$ in the norm $\|\cdot\|_{\mathrm{full}}$ is denoted $A \rtimes G$ or $C^*(G, A)$ and is called the *full crossed product of* A by G.

Observe that, letting G act trivially on $\mathbb{C}$, we have $\mathbb{C} \rtimes G \simeq C^*(G)$.

To a G-algebra A, we can associate a natural representation on a right Hilbert A-module. Consider the vector space

$$L^2(G,A) := \left\{ \varphi : G \to A : \int_G \varphi(s)^* \varphi(s) \mathrm{d}\mu(s) < \infty \right\},$$

where convergence of the integral is in the C^*-algebra A. In particular, note that the convergence condition is strictly weaker than the requirement that

$$\int_G \|\varphi(g)\|_A^2 \mathrm{d}\mu(s) < \infty.$$

The right A-module $L^2(G,A)$ is a Hilbert C^*-module over A in the inner product

$$\langle \varphi, \psi \rangle_A := \int_G \varphi(s)^* \psi(s) \mathrm{d}\mu(s).$$

There is a covariant representation of (A, α) on $L^2(G,A)$ given by

$$u : G \to \mathbb{U}(L^2(G,A)), \quad u_s(\varphi)(t) := \varphi(s^{-1}t)$$

$$\pi : A \to \mathrm{End}_A^*(L^2(G,A)), \quad \pi(a)\varphi(t) := \alpha_{t^{-1}}(a)\varphi(t),$$

giving a $*$-homomorphism $\pi \rtimes u : C_c(G,A) \to \mathrm{End}_A^*(L^2(G,A))$.

Definition 1.3.20. The *reduced C^*-norm* on $C_c(G,A)$ is

$$\|f\|_{\mathrm{red}} := \|\pi \rtimes u(f)\|_{\mathrm{End}_A^*(L^2(G,A))}.$$

The *reduced crossed product of A by G* is the completion of $C_c(G,A)$ in the norm $\|\cdot\|_{\mathrm{red}}$ and is denoted $A \rtimes_r G$ or $C_r^*(G,A)$.

As with the full crossed product, the trivial action of G on $\mathbb{C}$ gives $C_r^*(G) = \mathbb{C} \rtimes_r G$. The identity map $\mathrm{Id} : C_c(G,A) \to C_c(G,A)$ extends to a surjective $*$-homomorphism

$$A \rtimes G \to A \rtimes_r G.$$

This map is not injective in general, already when $A = \mathbb{C}$. The map $C^*(G) \to C_r^*(G)$ is an isomorphism if and only if the group G is *amenable*.

1.3.21 Free and Proper Actions

Let X be a locally compact Hausdorff space and G a locally compact group acting on X freely and properly. The orbit space X/G is a locally compact Hausdorff space by virtue of the nice properties of the action. Consequently, we are presented with two natural C^*-algebras associated with this action: the function algebra $C_0(X/G)$ and the crossed product $C_0(X) \rtimes G$.

The function space $C_c(X)$ comes equipped with a conditional expectation $\rho : C_c(X) \to C_c(X/G)$ defined by integration over the fibers of the quotient map $\pi : X \to X/G, x \mapsto [x]$:

$$\rho(f)([x]) := \int_{\pi^{-1}(x)} f(y)\mathrm{d}y = \int_G f(xg)\mathrm{d}g.$$

Using this map, we form the full Hilbert C^*-module $\mathscr{L}(X)$ as the completion of $C_c(X)$ in the norm induced by the $C_0(X/G)$-valued inner product

$$\langle f_1, f_2\rangle([x]) = \int_G \overline{f_1(xg)} f_2(xg)\mathrm{d}g.$$

By construction, the group G acts by unitaries on the module $\mathscr{L}(X)$ via $u_g(f)(x) := f(xg)$, and the function algebra $C_0(X)$ acts by pointwise multiplication $\pi(\varphi)(f)(x) := \varphi(x)f(x)$. This gives a covariant representation $(\mathscr{L}(X), \pi, u)$ of the G-C^*-algebra $C_0(X)$. The crossed product algebra $C_0(X) \rtimes G$ thus acts on $\mathscr{L}(X)$ through the integrated form $\pi \rtimes u$.

Theorem 1.3.22. *Let X be a locally compact Hausdorff space and G a locally compact group acting on X freely and properly. The covariant representation $(\mathscr{L}(X), \pi, u)$ induces an isomorphism*

$$\pi \rtimes u : C_0(X) \rtimes G \to \mathbb{K}(\mathscr{L}(X)).$$

Consequently, the C^*-algebras $C_0(X/G)$ and $C_0(X) \rtimes G$ are strongly Morita equivalent.

It is worth pointing out that in the special case where $X = G$, we obtain the fact that $C_0(G) \rtimes G$ is isomorphic to the algebra of compact operators on $L^2(G)$. This is equivalent to the Stone-von Neumann theorem ([23, Appendix C.6]).

From a broader perspective, this theorem motivates the viewpoint that for a general group action, the crossed product $C_0(X) \rtimes G$, which is always defined, is a substitute for the possibly ill-behaved quotient space X/G; a point of departure for the philosophy of noncommutative geometry.

A closed subgroup H of a locally compact group G acts freely and properly on G (viewed as a topological space). In this context, an extension of Theorem 1.3.22 to the case of two such subgroups is the following well-known result of Rieffel (see [26]).

Theorem 1.3.23. *Let G be a locally compact group and $H, K \subset G$ closed subgroups. The crossed product C^*-algebras $H \ltimes C_0(G/K)$ and $C_0(H \backslash G) \rtimes K$ are strongly Morita equivalent.*

For more on strong Morita equivalence and group actions on C^*-algebras, we recommend Rieffel's survey articles [27, 28].

1.4 C^*-Correspondences in Representation Theory

Let A and B be two C^*-algebras and (X, α) an (A, B)-correspondence. Using the $*$-homomorphism α and X, we can induce a Hilbert space representation $\pi : B \to \mathbb{B}(V_\pi)$ of B to one of A as follows. We consider the Hilbert space

$$X \otimes_B V_\pi$$

defined by the C^*-module tensor product mechanism we introduced earlier (see Definition 1.3.13). The action

$$a(x \otimes v) := \alpha(a)(x) \otimes v$$

of A on the space $X \otimes V_\pi$ gives rise to representation of A on the Hilbert space $X \otimes_B V_\pi$, which we will denote $\mathrm{Ind}_B^A(X, \pi)$ and refer to as the A-representation *induced from π via X.*

This induction procedure gives us a functor

$$\mathrm{Ind}_B^A(X) : \mathrm{Rep}(B) \to \mathrm{Rep}(A), \quad \pi \mapsto \mathrm{Ind}_B^A(X, \pi)$$

(see, e.g., [23, Proposition 2.69]) from the category $\mathrm{Rep}(B)$ of nondegenerate representations of B with bounded intertwining operators to the corresponding category $\mathrm{Rep}(A)$ of A. It respects unitary equivalence and direct sums. Moreover, if one equips both categories with the Fell topology (which is based on the notion of weak containment), then the earlier induction functor is continuous.

Let G and H be locally compact groups and let X be a Hilbert C^*-module over $C^*(H)$. Assume that there is a strongly continuous group homomorphism $\alpha : G \to \mathbb{U}(\mathrm{End}^*_{C^*(H)}(X))$. Just like in the case of a unitary representation on a Hilbert space, we can integrate to a homomorphism α : $C^*(G) \to \mathrm{End}^*_{C^*(H)}(X)$ (see, e.g., [23, Proposition C.17]) so that X becomes a $(C^*(G), C^*(H))$-correspondence. Now, the induction construction earlier gives us a functor

$$\mathrm{Ind}_{C^*(H)}^{C^*(G)}(X) : \mathrm{Rep}(C^*(H)) \longrightarrow \mathrm{Rep}(C^*(G)).$$

Recalling the categorical equivalence between the unitary representations of a group and representations of its C^*-algebra that we discussed earlier, we can view X as giving a functor

$$\mathrm{Ind}_H^G(X) : \mathrm{URep}(H) \longrightarrow \mathrm{URep}(G) \tag{1.4.0.1}$$

for the categories of unitary representations.

1.4.1 Proper Actions Revisited

We will revisit the material in Section 1.3.21 from a C^*-correspondence viewpoint and argue that Mackey's induction functor for unitary representations can be captured in the setting of group C^*-algebras via a suitable C^*-correspondence, as done by Rieffel originally.

Let X be a locally compact Hausdorff space and H a locally compact group acting on X properly. We do not assume that the action is free; however, we make the mild assumption that the locally compact space X/H is *paracompact*. Let λ denote the natural left action of G induced on $L^2(X)$: $\lambda(h)(f)(x) = f(h^{-1}x)$. We can turn $C_c(X)$ into a right $C^*(H)$-module via the formula

$$f \cdot b := \int_H b(h)\lambda(h^{-1})(f), \qquad f \in C_c(X),\ b \in C_c(H).$$

It can be shown that the completion $\mathscr{L}(X)$ of $C_c(X)$ with respect to the norm coming from the $C^*(H)$-valued form

$$\langle f, g\rangle_H(h) := \langle \lambda(h)(f), g\rangle_{L^2(X)}, \qquad f, g \in C_c(X),\ b \in C_c(H),\ h \in H$$

has the structure of a right Hilbert C^*-module over $C^*(H)$. Details can be found, for example, in [5, Section 2]. In particular, the paracompactness assumption is used for proving the positivity of the inner product. If the action is free, one can also show that the Hilbert C^*-module that we obtain is full.

We point out that if (σ, V_σ) is a unitary representation of H, then we have an isometry of Hilbert spaces

$$\mathscr{L}(X)\otimes_{C^*(H)} V_\sigma \to L^2_H(X, V_\sigma), \qquad [f\otimes v] \mapsto \left(x \mapsto \int_H f(h^{-1}x)\sigma(h)(v)dh\right) \tag{1.4.1.1}$$

where $L^2_H(X, V_\sigma)$ denotes the Hilbert space of L^2-sections of the bundle $X \times_H V_\sigma$ over X/H.

1.4.2 Induction of Unitary Representations

Now consider a locally compact Hausdorff group G with a closed subgroup H. Applying the construction of Section 1.4.1 with $X = G$, we obtain a full right

Hilbert C^*-module $\mathscr{L}(G)$ over $C^*(H)$. The left regular action of G on $C_c(G)$ leads to an action of $C^*(G)$ on $\mathscr{L}(X)$ by adjointable operators on the left, and thus $\mathscr{L}(G)$ becomes a $(C^*(G), C^*(H))$-correspondence.

Theorem 1.4.3 (Rieffel [25]). *The induction functor associated to the C^*-correspondence $\mathscr{L}(G)$ (see (1.4.0.1)) coincides with the induction functor of Mackey. That is, for any unitary representation σ of H, we have*

$$\mathrm{Ind}_H^G(\mathscr{L}(G))(\sigma) = \mathrm{Ind}_H^G(\sigma).$$

Moreover, we have

$$C_0(G/H) \rtimes G \simeq \mathbb{K}(\mathscr{L}(G)).$$

The first part of the theorem tells us that we can capture the induction of unitary representations via the framework of C^*-correspondences. Note that the second part of the theorem tells us that $C^*(H)$ and $C_0(G/H) \rtimes G$ are strongly Morita equivalent, which is an elegant C^*-algebraic formulation of the Imprimitivity Theorem of Mackey.

Before we move on to our next application, we should point out related works [5, 6, 24] in which *parabolic induction* (a fundamental tool in the representation theory of semisimple Lie groups, see Hochs' chapter) is approached via C^*-correspondence machinery.

1.4.4 Local Theta Correspondence

Our second representation theoretic application of C^*-correspondences is to the theory of *local theta correspondence*, a major theme in the theory of automorphic forms and representation theory.

In a nutshell, local theta correspondence, developed by Roger Howe in the mid-1970s, establishes a bijection between certain sets of admissible[1] irreducible representations of reductive groups G and H that form a *dual pair*, that is, G and H sit inside a large enough symplectic group in such a way that they form each other's centralizers. Roughly speaking, this bijection is obtained by considering how the so-called *oscillator representation*[2] (see, e.g., [19]) of the ambient symplectic group decomposes as a $G \times H$-representation.

In [18] (see also [13, 30]), it is proven that when one group, say H, is much smaller than the other (this is the so-called *stable range* setting), remarkable

[1] Admissibility is a type of finiteness condition. The category of admissible representations contains that of unitary ones.

[2] Also known as the metaplectic representation or Shale–Weil representation.

things happen: theta correspondence preserves unitarity, all of the unitary dual of H enters the bijection so that we get an embedding

$$\theta : \widehat{H} \hookrightarrow \widehat{G} \tag{1.4.4.1}$$

and the theta lift of an irreducible unitary representation of the small group can be *explicitly* described via a construction that involves integrating matrix coefficients.

The next result, proven in [21], says all of the earlier remarkable features of the stable range setting can be explained by the single fact that there is a C^*-correspondence at work behind the scenes.

Theorem 1.4.5. *Let (G,H) be a Type I dual pair in the stable range with H being the smaller group. The injection (1.4.4.1) above given by local theta correspondence is induced by the induction functor associated with a $(C^*(G),C^*(H))$-correspondence.*

More precisely, one can construct a $(C^*(G),C^*(H))$-correspondence Θ from the oscillator representation with the property that the induction functor

$$\mathrm{Ind}_H^G(\Theta) : \mathrm{URep}(H) \longrightarrow \mathrm{URep}(G)$$

satisfies

$$\mathrm{Ind}_H^G(\Theta)(\sigma^*) \simeq \theta(\sigma)$$

for any $\sigma \in \widehat{H}$. Here, σ^* is the contragradient of σ. Note that, as a consequence, the embedding (1.4.4.1) is continuous.

Let us give an indication of how Θ is constructed. To ease the discussion, we consider the following simple special case: Let G denote the symplectic group $\mathrm{Sp}_{2n}(\mathbb{R})$ and H denote the orthogonal group $O(p,q)$ with $n = p+q$ and n even. We begin by considering the oscillator representation

$$\omega : G \times H \to \mathcal{U}(L^2(M_n(\mathbb{R}))).$$

The orthogonal group acts on $L^2(M_n(\mathbb{R}))$ (here $M_n(\mathbb{R})$ denotes the space of $n \times n$ real matrices) by left multiplication, so that on the dense open subset $\mathrm{GL}_n(\mathbb{R})$, the action of H is free and proper. This is the key for our C^*-algebraic take on the situation. Indeed, following the general construction in Section 1.4.1 with $X = \mathrm{GL}_n(\mathbb{R})$, one constructs a full Hilbert C^*-module Θ over $C^*(H)$. It can be shown that Θ contains the Schwartz space $\mathcal{S}(M_n(\mathbb{R}))$ (which coincides with the space of smooth vectors in the oscillator representation ω) as a dense submodule. It follows that the module Θ, again via the oscillator representation, admits an adjointable action of $C^*(G)$ on the left, giving us our $(C^*(G),C^*(H))$-correspondence.

1.5 Fundamentals of K-Theory

By the Gelfand–Naimark theorem, C^*-algebras are generalization of locally compact Hausdorff topological spaces. The philosophy of noncommutative geometry is to view a noncommutative C^*-algebra as the algebra of functions on a nonexistent topological space. As such, algebraic topology for C^*-algebras amounts to studying homotopy invariants of such spaces. Standard references for K-theory are [3, 29].

We consider the category C^*, whose objects are C^*-algebras and morphisms are $*$-homomorphisms, as well as the category Ab of abelian groups and group homomorphisms. Topological K-theory is a functor $K_* : C^* \to \mathrm{Ab}$ with certain exactness properties.

1.5.1 Elementary Description via Projective Modules

There are several different descriptions of K-theory. We start with the simplest of these, which is also closest to the origin of K-theory in algebra. In this section, A will be a C^*-algebra with unit.

A finitely generated right A-module P is projective if there exists $n \in \mathbb{N}$ and a right A-module Q such that $P \oplus Q \simeq A^n$. Equivalently, if M is a right A module and $q : M \to P$ is a surjective A-module map, then there exists an A module map $s : P \to M$ such that $q \circ s = \mathrm{Id}_P$.

We denote the isomorphism class of a projective module P by $[P]$ and define

$$K_0(A) := \{[P] - [Q] : P, Q \text{ finitely generated projective } A\text{-modules}\},$$

the group of formal differences of isomorphism classes of finitely generated projective A-modules, subject to the relation $[P \oplus Q] = [P] + [Q]$. The set $K_0(A)$ is an abelian group by construction, with 0 element given by the 0 module.

An equivalent description of the K-theory group of a unital C^*-algebra A is obtained by considering projections $p \in M_n(A)$, $p = p^2 = p^*$. In order to be able to work with projections of arbitrary size, we consider the maps

$$M_n(A) \to M_{n+1}(A), \quad T \mapsto \begin{pmatrix} T & 0 \\ 0 & 0 \end{pmatrix}.$$

Using these maps, we can consider any two projections to be in the same matrix algebra. Alternatively, we can work with projections in the algebra $M_\infty(A)$ of matrices of arbitrary finite size. Two projections p and q are *equivalent* if there is an invertible element u such that $p = u^{-1}qu$. The set of equivalence classes of projections is not a group in general, but only a semigroup under the operation induced by the direct sum of matrices:

$$M_n(A) \times M_k(A) \ni (p,q) \mapsto p \oplus q := \begin{pmatrix} p & 0 \\ 0 & q \end{pmatrix}.$$

This operation maps projections to projections and is well defined on equivalence classes. That is, one can add the equivalence classes of projections p and p as $[p] + [q] = [p \oplus q]$. To get a group, one has to add inverses in a formal way. Therefore, we define

$$K_0(A) := \{[p] - [q] : p, q \text{ projections in } M_\infty(A), [p \oplus q] = [p] + [q]\}$$

which is a group by definition.

Given a projection $p \in M_n(A)$, we obtain a finitely generated projective module by setting $P := pA^n$. This correspondence defines an isomorphism between the two definitions of $K_0(A)$ that we have given earlier.

1.5.2 Elementary Description via Unitaries

A second group can be constructed using unitaries. Let A be a unital C^*-algebra and denote by $\mathbb{U}_n(A)$ the unitary group of the matrix algebra $M_N(A)$. In terms of Hilbert C^*-modules, we have $\mathbb{U}_n(A) = \mathbb{U}(A^n)$, the group of adjointable unitary transformations of the module A^n. As with projections, there are maps

$$\mathbb{U}_n(A) \to \mathbb{U}_{n+1}(A), \quad u \mapsto \begin{pmatrix} u & 0 \\ 0 & 1 \end{pmatrix},$$

and we write $\mathbb{U}(A) := \lim_{n\to\infty} \mathbb{U}_n(A)$. In order to construct our groups, we recall the C^*-version of the Whitehead Lemma.

Lemma 1.5.3. *Let $u, v \in \mathbb{U}_n(A)$ be a unitaries. Then there is a continuous path of unitaries*

$$\gamma : [0,1] \to \mathbb{U}_{2n}(A), \quad \gamma(0) = \begin{pmatrix} u & 0 \\ 0 & v \end{pmatrix}, \quad \gamma(1) = \begin{pmatrix} uv & 0 \\ 0 & 1_n \end{pmatrix},$$

where $1_n \in M_n(\mathbb{C})$ denotes the identity.

Proof. The map

$$U : [0,1] \to M_{2n}(\mathbb{C}), \quad U(t) := \begin{pmatrix} 1_n \cos t & 1_n \sin t \\ -1_n \sin t & 1_n \cos t \end{pmatrix},$$

is a continuous path of unitaries connecting 1_{2n} to $\begin{pmatrix} 0 & 1_n \\ -1_n & 0 \end{pmatrix}$. Now write

$$\begin{pmatrix} u & 0 \\ 0 & v \end{pmatrix} = \begin{pmatrix} u & 0 \\ 0 & 1_n \end{pmatrix} \begin{pmatrix} 0 & -1_n \\ 1_n & 0 \end{pmatrix} \begin{pmatrix} v & 0 \\ 0 & 1_n \end{pmatrix} \begin{pmatrix} 0 & 1_n \\ -1_n & 0 \end{pmatrix},$$

which is homotopic to $\begin{pmatrix} uv & 0 \\ 0 & 1_n \end{pmatrix}$, by using $U(t)$ and $U(t)^*$ on the relevant factors. □

Two unitaries $u \in \mathbb{U}_n(A)$ and $v \in \mathbb{U}_k(A)$ are *homotopic* if they can be connected by a continuous path of unitaries in $\mathbb{U}(A)$. We define the *sum* of two such unitaries u and v as

$$u \oplus v := \begin{pmatrix} u & 0 \\ 0 & v \end{pmatrix} \in \mathbb{U}_{n+k}(A).$$

By the Whitehead Lemma, there is a homotopy

$$\gamma : [0,1] \to \mathbb{U}_{n+k}(A), \quad \gamma(0) = \begin{pmatrix} u & 0 \\ 0 & v \end{pmatrix}, \quad \gamma(1) = \begin{pmatrix} uv & 0 \\ 0 & 1 \end{pmatrix}.$$

Therefore, the set $\pi_0(\mathbb{U}(A))$ of homotopy classes of unitaries is an abelian group.

Definition 1.5.4. Let A be a unital C^*-algebra. The *odd K-theory group* of A is

$$K_1(A) := \pi_0(\mathbb{U}(A)) = \{[u] : \exists n \in \mathbb{N},\ u \in \mathbb{U}_n(A)\}$$

The association $A \mapsto K_1(A)$ is functorial for $*$-homomorphisms $\varphi : A \to B$ by simply setting $\phi_*[u] = [\varphi(u)]$.

1.5.5 Vector Bundles and Topological K-Theory

For a compact Hausdorff space X, we obtain two K-theory groups

$$K^*(X) := K_*(C(X)),$$

which are *contravariant* functors when defined on the category of compact Hausdorff spaces.

Via the Serre–Swan theorem, these groups admit a purely topological description. In the case of K_0, this is quite straightforward: Any finitely generated projective module over $C(X)$ arises (up to isomorphism) as the module of sections of some locally trivial finite dimensional vector bundle $E \to X$. Two such bundles E and F are said to be *isomorphic* if there exists a homeomorphism $g : E \xrightarrow{\sim} F$ such that $\pi_F \circ g = \pi_E$, and g restricts to an isomorphism of vector spaces in each fiber. In other words, $E \simeq F$ as vector bundles if and only if $\Gamma(E) \simeq \Gamma(F)$ as finitely generated projective $C(X)$-modules.

We denote by $[E]$ the isomorphism class of a vector bundle $E \to X$ and by Vect(X) the set of isomorphism classes of locally trivial finite dimensional vector bundles over X. This leads us to the description

$$K^0(X) := \{[E] - [F] : [E],[F] \in \mathrm{Vect}(X),\}$$

on which we impose the relation $[E] + [F] = [E \oplus F]$ where

$$E \oplus F := \{(e,f) \in E \times F : \pi_E(e) = \pi_F(f)\}$$

denotes the *Whitney sum* of the bundles. It is a straightforward verification that the Whitney sum induces an isomorphism of projective modules

$$\Gamma(E \oplus F) \simeq \Gamma(E) \oplus \Gamma(F).$$

For $K^1(X)$, a topological description is given by considering triples (E,F,g) where E and F are vector bundles over X and g is a vector bundle isomorphism. Two such triples are equivalent if there is a homotopy between the isomorphisms g and g'.

1.5.6 K-Theory for Nonunital Algebras

Much of the significance and computability of K-theory come from the existence of exact sequences associated with extensions

$$0 \to I \xrightarrow{i} A \xrightarrow{\pi} B \to 0$$

of C^*-algebras. In such an extension, $i : I \to A$ is the inclusion of a closed two-sided ideal of A and $\pi : A \to B$ is a surjective $*$-homomorphism such that $\ker \pi = I$. In topology, such exact sequences arise from the inclusion of an open set $U \subset X$ into a locally compact Hausdorff X and considering

$$0 \to C_0(U) \to C_0(X) \to C_0(X \setminus U) \to 0,$$

where $C_0(X)$ denotes the functions vanishing at infinity on X.

Note that an ideal $I \subset A$ cannot contain the unit of the algebra A, so in order to associate K-groups to I, we need to extend our definition.

Definition 1.5.7. Let A be a C^*-algebra, viewed as a Hilbert C^*-module over itself as in Example 1.3.6. The *multiplier algebra of* A is the C^*-algebra $\mathrm{End}^*_A(A)$ of adjointable operators on A.

By definition, $M(A)$ is a unital C^*-algebra and since $\mathbb{K}(A) = A$, $A \subset M(A)$ is a closed two-sided essential ideal of $M(A)$. The multiplier algebra is the largest such C^*-algebra and has the following universal property: If B is a unital C^*-algebra containing A as an essential ideal, then there is a unique $*$-homomorphism $\phi : B \to M(A)$ such that $\phi|_A = \mathrm{Id}$. In case A is unital, we have $M(A) = A$, but in general, $M(A)$ is a nonseparable C^*-algebra (even if A itself is separable). For instance, if X is a locally compact Hausdorff space, then $M(C_0(X)) = C(X^\beta)$ where X^β is the Stone–Čech compactification of X.

The earlier discussion shows that we can view the multiplier algebra as the "maximal" unitization of A.

Definition 1.5.8. The *(minimal) unitization of* A is the C^*-subalgebra $A^+ := C^*(A, 1)$ of $M(A)$ generated by $1 \in M(A)$ and $A \subset M(A)$.

There is an intrinsic algebraic construction of A^+ (for nonunital A) as the vector space $A^+ := A \oplus \mathbb{C}$ equipped with coordinatewise addition and multiplication

$$(a_1, \lambda_1) \cdot (a_2, \lambda_2) := (a_1 a_2 + \lambda_1 a_2 + \lambda_2 a_1, \lambda_1 \lambda_2).$$

This yields an isomorphic complex algebra. The advantage of Definition 1.5.8 is that A^+ is automatically a C^*-algebra, and that $A^+ = A$ if A already has a unit. In either case, A is a closed two-sided ideal in A^+. In case A is nonunital, there is a surjective $*$-homomorphism

$$\pi^+ : A^+ \to \mathbb{C}, \quad (a, \lambda) \mapsto \lambda$$

with kernel A. Since K_0 and K_1 as constructed for unital C^*-algebras are functorial for $*$-homomorphisms, we arrive at Definition 1.5.9.

Definition 1.5.9. Let A be a nonunital C^*-algebra. The K-groups of A are

$$K_*(A) := \ker(\pi^+_* : K_*(A^+) \to K_*(A)).$$

Equivalently, we have the explicit descriptions

$$K_0(A) := \{[p] - [q] \in K_0(A^+) : [\pi^+(p)] = [\pi^+(q)]\}$$
$$K_1(A) := \{[u] \in K_1(A^+) : [\pi^+(u)] = [1]\}.$$

It is straightforward to see that $A \to K_*(A)$ is functorial for general (not necessarily unital) $*$-homomorphisms. Given an extension

$$0 \to I \xrightarrow{i} A \xrightarrow{\pi} B \to 0, \tag{1.5.9.1}$$

there is an exact sequence

$$K_1(I) \xrightarrow{i_*} K_1(A) \xrightarrow{\pi_*} K_1(B) \xrightarrow{\partial} K_0(I) \xrightarrow{i_*} K_0(A) \xrightarrow{\pi_*} K_0(B). \qquad (1.5.9.2)$$

The map $K_1(B) \xrightarrow{\partial} K_0(I)$ is often called the *index map*.

Using the extended definition of K-theory, we can state an additivity property for infinite direct sums. Let $\{A_i\}_{i\in I}$ be a family of C^*-algebras and consider the algebra

$$\bigoplus_{i\in I}^{\text{alg}} A_i := \left\{ f : I \to \bigsqcup_{i\in I} A_i : f(i) \in A_i,\ f(i) = 0 \text{ for all but finitely many } i \right\},$$

with pointwise product $(fg)(i) := f(i)g(i)$ and the norm $\|f\| := \sup_{i\in I} \|f(i)\|$. The so obtained completion is denoted $\bigoplus_{i\in I} A_i$ and called the C_0*-direct sum* of the family $\{A_i\}_{i\in I}$. Note that for infinite index sets I, the C_0-direct sum yields a nonunital C^*-algebra, even if each A_i has a unit. For abelian groups, the direct sum is defined in the usual algebraic manner, and K-theory is compatible with this notion of direct sum.

Proposition 1.5.10. *Let $\{A_i\}_{i\in I}$ be a family of C^*-algebras. The map*

$$K_0\left(\bigoplus_{i\in I} A_i\right) \to \bigoplus_{i\in I} K_0(A_i), \quad [p] - [q] \mapsto ([p(i)] - [q(i)])_{i\in I}$$

is a well-defined isomorphism of K-groups.

A remark on the well-definedness of the above map. If $f \in \bigoplus_{i\in I} A_i$, then for every $\varepsilon > 0$, we have $\|f(i)\| < \varepsilon$ for all but finitely many i. Now if $p : I \to M_n\left(\bigoplus_{i\in I} A_i\right)^+$ is projection valued, then for each i we have that $f(i)$ is a projection in $M_n(A_i^+)$. Thus, $\|f(i)\| \in \{0, 1\}$ for all i, which implies that $f(i) = 0$ for all but finitely many i. That is, projections are always defined over the algebraic direct sum in the sense that they all come from maps $p : I \to M_n\left(\bigoplus_{i\in I}^{\text{alg}} A_i\right)^+$.

1.5.11 Higher K-Theory and Bott Periodicity

For a given C^*-algebra, we define its *cone* CA and *suspension* SA by

$$CA := C_0((0, 1], A), \quad SA := C_0((0, 1), A),$$

both equipped with the sup norm of A-valued functions. Evaluation at 1 induces a surjective $*$-homomorphism

$$\text{ev}_1 : CA \to A, \quad f \mapsto f(1),$$

with kernel $\ker \mathrm{ev}_1 = SA$, so we have an extension

$$0 \to SA \to CA \to A \to 0. \tag{1.5.11.1}$$

The C^*-algebra CA is *contractible*: There is a continuous family of $*$-homomorphisms

$$\phi_s : CA \to CA, \quad \phi_s(f)(t) := f(st) \quad s \in [0,1],$$

such that $\phi_0 = 0$ and $\phi_1 = \mathrm{Id}_{CA}$. Homotopy invariance of K-theory now implies that $\mathrm{Id}_* : K_*(CA) \to K_*(CA)$ induces the 0 map, which is possible only if $K_*(CA)$ is trivial. From the exactness of the sequence (1.5.9.2), we conclude:

Lemma 1.5.12. *The index map associated with the cone extension* (1.5.11.1) *induces a functorial isomorphism* $K_1(A) \xrightarrow{\sim} K_0(SA)$.

The suspension operation $A \to SA$ is a functor on the category of C^*-algebras. Since we have $K_1 = K_0 \circ S$, we arrive at the following:

Definition 1.5.13. The higher K-groups $K_n(A)$ for a C^*-algebras are defined to be $K_n(A) := K_0(S^n A)$.

With this definition of higher K-theory, we deduce the existence of a long exact sequence extending (1.5.9.2). For an extension (1.5.9.1), we have a (left infinite) long exact sequence

$$\cdot \to K_n(I) \to K_n(A) \to K_n(B) \to K_{n-1}(I) \to \cdots \to K_0(A) \to K_0(B)$$

of K-groups. One of the most important results in K-theory is *Bott periodicity*, which states that the operation of taking suspensions is two-periodic at the level of K-theory.

Theorem 1.5.14. *Let A be a C^*-algebra. There is a functorial isomorphism $K_0(S^2A) \simeq K_0(A)$ and consequently $K_{n+2}(A) \simeq K_n(A)$.*

An important consequence of Bott periodicity is the fact that the exact sequence (1.5.9.2) is in fact finite and cyclic.

Corollary 1.5.15. *An extension of C^*-algebras*

$$0 \to I \to A \to B \to 0$$

induces a periodic six-term exact sequence

$$\begin{array}{ccccc} K_0(I) & \longrightarrow & K_0(A) & \longrightarrow & K_0(B) \\ \uparrow & & & & \downarrow \\ K_1(B) & \longleftarrow & K_1(A) & \longleftarrow & K_1(I) \end{array} \tag{1.5.15.1}$$

of K-groups.

The cyclic six-term exact sequence is the main tool for the computation of K-groups in concrete cases.

1.5.16 Kasparov Picture and Morita Invariance

Based on the theory of Fredholm operators, we obtain another description of K-theory. This description is closely related to Kasparov's KK-theory [15], a generalization of K-theory to a bivariant setting which we will not describe here. The Kasparov picture is nonetheless useful for our setting; in particular, it offers a neat perspective on the Morita invariance of K-theory.

Definition 1.5.17. Let A be a C^*-algebra. A *Kasparov triple* (X^0, X^1, F) for A consists of two countably generated Hilbert C^*-modules X^i over A together with an operator $F : X^0 \to X^1$ such that

$$1 - FF^* \in \mathbb{K}(X^1, X^0), \quad 1 - F^*F \in \mathbb{K}(X^0, X^1).$$

A *homotopy* of Kasparov triples (X_i^0, X_i^1, F_i), $i = 0, 1$ is a Kasparov triple (Z^0, Z^1, F_Z) over $C([0,1], A)$ such that

$$\mathrm{ev}_0(Z^0 \otimes_{\mathrm{ev}_i} A, Z^1 \otimes_{\mathrm{ev}_i} A, F_Z \otimes_{\mathrm{ev}_i} 1) = (X_i^0, X_i^1, F_i).$$

It can be shown that such homotopies define an equivalence relation and that the set of equivalence classes of Kasparov triples modulo homotopy form an abelian group, which Kasparov denotes $KK_0(\mathbb{C}, A)$.

Theorem 1.5.18. *Let A be a C^*-algebra. The map*

$$K_0(A) \to KK_0(\mathbb{C}, A), \quad [p] - [q] \mapsto (pA^n, qA^n, 0) \tag{1.5.18.1}$$

is a well-defined isomorphism of abelian groups.

In case the operator $F : X^0 \to X^1$ has a closed range, both $\ker F \subset X^0$ and $\ker F^* \subset X^1$ are complemented submodules. In this case, the inverse to the map (1.5.18.1) is given by $(X^0, X^1, F) \mapsto [\ker F] - [\ker F^*]$, which is a module operator analogue of the Fredholm index.

Given a Hilbert C^*-module Y_B and a $*$-homomorphism $A \to \mathbb{K}(Y)$, we obtain a bimodule ${}_AY_B$. We can use the C^*-module tensor product (see 1.3.13) to transform modules over A into modules over B via

$$X_A \mapsto X \otimes_A Y,$$

and we can transfer operators on X to operators $X \otimes_A Y$ via $T \mapsto T \otimes 1$. Combining these allows us to construct a mapping on the level of Kasparov triples.

Proposition 1.5.19. *Let Y_B be a Hilbert C^*-module and $A \to \mathbb{K}(Y)$ a $*$-homomorphism. If $T \in \mathbb{K}(X)$, then $T \otimes 1 \in \mathbb{K}(X \otimes_A Y)$. In particular, the map*

$$- \otimes_A Y : KK_0(\mathbb{C}, A) \to K_0(\mathbb{C}, B), \quad (X^0, X^1, F) \mapsto (X^0 \otimes_A Y, X^1 \otimes_A Y, F \otimes 1)$$

is well-defined. If ${}_AY_B$ is a Morita equivalence bimodule, then $- \otimes_A Y$ is an isomorphism.

The first part of the proposition is proved in [17, Proposition 4.7]. The subsequent parts follow since

$$1 - (F \otimes 1)^*(F \otimes 1) = (1 - F^*F) \otimes 1, \quad 1 - (F \otimes 1)(F \otimes 1)^*,$$

giving the well-definedness of the map $- \otimes_A Y$ on the Kasparov groups. If Y is a Morita equivalence, then we have isomorphisms $Y \otimes_B Y^* \simeq A$ and $Y^* \otimes_A Y \simeq B$, which amount to the statement that

$$(- \otimes_A Y) \circ (- \otimes_B Y^*) = (- \otimes_A Y \otimes_B Y^*) = (- \otimes_A A) = \mathrm{Id},$$

and similarly $(- \otimes_B Y^*) \circ (- \otimes_A Y) = \mathrm{Id}$.

Corollary 1.5.20. *Let A and B be Morita equivalent C^*-algebras. There are functorial isomorphisms $K_*(A) \to K_*(B)$.*

This corollary is derived using the isomorphisms $KK_0(\mathbb{C}, A) \to K_0(A)$ and $K_1(A) \simeq K_0(SA)$.

1.6 K-Theory of Group C^*-Algebras

As mentioned earlier, in the philosophy of noncommutative geometry, the C^*-algebra $C^*(G)$ of a locally compact Hausdorff group G is viewed as the algebra of functions on the, possibly non-Hausdorff, unitary dual $\widehat{G}$ of G. So the basic perspective is that the K-theory of $C^*(G)$ "captures topological information" of $\widehat{G}$.

Describing the K-theory of $C^*(G)$ in general is a very difficult problem. In the case of reductive linear groups over local fields, one can use deep results from representation theory that "classify" tempered representations to obtain a description of $C_r^*(G)$. This approach is discussed at length in the chapters of Aubert and Hochs. More generally, the celebrated Baum–Connes conjecture (see [2]) offers a solution to this problem through geometry. Specializing this conjecture to the case of semisimple Lie groups, one finds a rich interplay of index theory, representation theory, and K-theory. These connections are the content of Wang's chapter. In this section, our goal will be very modest: We will take a look at the part of K-theory of $C^*(G)$ for a liminal group G that arises from the isolated points of $\widehat{G}$.

1.6.1 Liminal Groups

We call a group G *liminal*[3] if $\pi(C^*(G))$ is equal to the compact operators $\mathbb{K}(\mathbb{V}_\pi)$ for every irreducible unitary representation (π, V_π) of G. For example, compact groups, nilpotent groups (Dixmier, Fell), and reductive groups (Harish-Chandra) are all known to be liminal.

Let us assume that G is a liminal separable group. Under this assumption, every point in $\widehat{G}$ is closed (see [7, 9.5.3]). We will be interested in the isolated, that is, open points of $\widehat{G}$. If π is an isolated point, then $\{\pi\}$ is both closed and open in $\widehat{G}$ and basic C^*-algebra theory implies that the closed two-sided ideal $\ker(\pi)$ of $C^*(G)$ is complemented:

$$C^*(G) \simeq \ker(\pi) \oplus J_\pi,$$

where J_π is the closed two-sided ideal

$$J_\pi := \bigcap_{\substack{\sigma \in \widehat{C^*(G)} \\ \sigma \neq \pi}} \ker(\sigma),$$

and the sum is a direct sum of C^*-algebras.

Lemma 1.6.2. *Let G be a liminal separable group. If $\{\pi\}$ is isolated in $\widehat{G}$, then π defines a class $[\pi] \in K_0(C^*(G))$ of infinite order.*

Proof. As G is liminal, $\pi : C^*(G) \to \mathbb{K}(V_\pi)$ is surjective. It follows that

$$J_\pi \simeq C^*(G)/\ker(\pi) \simeq \mathbb{K}(V_\pi).$$

As J_π is a direct summand, the injection $J_\pi \hookrightarrow C^*(G)$ leads to an injection

$$K_0(J_\pi) \to K_0(C^*(G)).$$

Since $K_0(\mathbb{K}(V_\pi)) \simeq \mathbb{Z}$, we conclude that there is a copy of $\mathbb{Z}$ in $K_0(C^*(G))$ that is contributed by π. □

[3] Also known as *CCR*.

1.6.3 Isolated Points and Integrability

As we have argued earlier, isolated points of $\widehat{G}$ are of importance for the K-theory of group C^*-algebras. Let us summarize some results regarding such points.

Recall that a unitary irreducible representation (π, V_π) of G is called *integrable* if for some nonzero vectors $v, w \in V_\pi$, the matrix coefficient function

$$g \mapsto \langle v, \pi(g)(w) \rangle \tag{1.6.3.1}$$

lies in $L^1(G)$.

Theorem 1.6.4 ([8, 32]). *Let G be unimodular and $\pi \in \widehat{G}$ be integrable. Then, π is isolated in $\widehat{G}$.*

When the matrix coefficients (1.6.3.1) lie in $L^2(G)$, we call π *square-integrable*. Such π may not be integrable, but for the reduced dual $\widehat{G}_{red}$ (see Def. 1.2.12 and the remarks after that), they play a similar role.

Theorem 1.6.5 ([12]). *Let G be an almost connected unimodular group. If $\pi \in \widehat{G}$ is square-integrable, then π is isolated in the reduced dual $\widehat{G}_{red}$.*

From the perspective of K-theory, the classes arising from (square-) integrable representations can also be accessed through the *minimal projections* that are obtained by suitably normalizing the matrix coefficients (1.6.3.1). A detailed discussion can be found in [31].

Finally, let us illustrate the earlier above theorem by considering the group $G = \mathrm{SL}_2(\mathbb{R})$. Consider the discrete series representations $D_n^\pm$ with $n > 0$ as defined in Ex. 2.3.38 of Hochs' chapter. Note that the limit of discrete series representations is denoted $D_0^\pm$. For $n > 1$, the $D_n^\pm$ are integrable and hence they lead to generators of $K_0(C^*(G))$. However, $D_1^\pm$ are *not* integrable, they are only square-integrable. In the unitary dual of G, together with the trivial representation, they lie at the end of the complementary series (see, e.g., Theorem 3.1 of [22]). This changes when one considers $C_r^*(G)$ and the reduced dual: $D_1^\pm$ are isolated in $\widehat{G}_{red}$ (the trivial representation and the complementary series are not in the reduced dual!) and lead to generators of K-theory of $C_r^*(G)$.

1.7 Equivariant KK-Theory and the Connes–Kasparov Conjecture

We finish our chapter with a short summary of equivariant KK-theory, the γ-element, and the Connes–Kasparov conjecture. We hope that this discussion will be a useful companion to the later parts of Wang's chapter.

1.7.1 G-Correspondences

Let G be a locally compact group and (A, B) a pair of G-C^*-algebras. We will enhance the class of C^*-correspondences of (A, B) by incorporating a strongly continuous G-action on them.

Definition 1.7.2. Let G be a locally compact group, (B, β) a G-C^*-algebra, and X a Hilbert C^*-module over B. A *strongly continuous action of G on X* is a group homomorphism

$$G \to \mathrm{Aut}_{\mathbb{C}} X, \quad g \mapsto \xi_g$$

into the continuous *linear* automorphisms on X, such that:

1. for all $x \in X$, the map
$$G \to X, \quad g \mapsto \xi_g(x)$$
is continuous for the norm topology on X;
2. the inner product on X satisfies
$$\langle \xi_g(x), \xi_g(y) \rangle = \beta_g(\langle x, y \rangle).$$

We refer to (X, ξ) as a C^*-G-module over B.

We emphasize that in case B is a nontrivial G-C^*-algebra, the action of G on the module X is *not* implemented by right B-linear module operators. In fact, this definition implies that for all $b \in B$, $x \in X$ and $g \in G$, we have $\xi_g(xb) = \xi_g(x)\beta_g(b)$, since

$$\langle \xi_g(xb) - \xi_g(x)\beta_g(b), \xi_g(xb) - \xi_g(x)\beta_g(b) \rangle = 0$$

by equivariance of the inner product.

Example 1.7.3. A key example comes from taking $X = \Omega^1(M)$, the module of 1-forms on a Riemannian manifold and $G = \mathrm{Isom}(M)$.

Given a C^*-G-B-module X, the algebra of compact operators $\mathbb{K}(X)$ becomes a G-C^*-algebra via

$$\alpha_g(T)(x) := \alpha_g T \alpha_{g^{-1}}(x), \quad g \in G, \quad x \in X, \quad T \in \mathbb{K}(X).$$

The strong continuity of the action of G on $\mathbb{K}(X)$ follows from the strong continuity of the action on X. However, the reader should be warned that this does not hold for the entire endomorphism algebra $\mathrm{End}^*_B(X)$. This fact leads us to the notion of a G-continuous operator: An element $T \in \mathrm{End}^*_B(X)$ is G-continuous if the map $g \mapsto \alpha_g(T)(x)$ is continuous for all $x \in X$.

Definition 1.7.4. Let (A,α) and (B,β) be G-C^*-algebras. An *equivariant C^*-correspondence* is an (A,B)-correspondence X such that X is a right C^*-G-module over B and

$$\xi_g(ax) = \alpha_g(a)\xi_g(x).$$

The case where B is equipped with the trivial G-action is of special interest, for then $\xi : G \to \mathrm{Aut}_{\mathbb{C}}(X)$ in fact defines a strongly continuous unitary representation of G on X. Together with the A representation $\pi : A \to \mathrm{End}^*_B(X)$, the triple (π,ξ,X) is a covariant representation of (A,α) in the sense of Definition 1.3.17.

1.7.5 Equivariant KK-Theory

The Kasparov picture of K-theory can be easily adapted to the G-equivariant setting [16]. In this section, we will also adopt the bivariant viewpoint that we have not elaborated on before.

Definition 1.7.6. Let (A,B) be a pair of G-C^*-algebras. An equivariant *Kasparov module* (X^0,X^1,F) for (A,B) consists of two countably generated C^*-G-correspondences X^i for (A,B) together with an operator $F : X^0 \to X^1$ such that

$$a(1-FF^*) \in \mathbb{K}(X^1,X^0), \quad Fa - aF^* \in \mathbb{K}(X^0,X^1), \quad a(F-\alpha_g(F)) \in \mathbb{K}(X),$$

and the latter defines a G-continuous map $g \mapsto a(F-\alpha_g(F))$. A *homotopy* of equivariant Kasparov modules (X^0_i,X^1_i,F_i), $i=0,1$ is an equivariant Kasparov module (Z^0,Z^1,F_Z) for $(A,C([0,1],B))$ such that

$$\mathrm{ev}_0(Z^0 \otimes_{\mathrm{ev}_i} B, Z^1 \otimes_{\mathrm{ev}_i} B, F_Z \otimes_{\mathrm{ev}_i} 1) = (X^0_i,X^1_i,F_i).$$

Homotopy of G-equivariant Kasparov (A,B)-modules defines an equivalence relation and the set of equivalence classes $KK^G_0(A,B)$ forms an abelian group with addition induced by the direct sum of Kasparov modules.

In case the group G is compact, $A=\mathbb{C}$ and $B=C(M)$ with M a compact Hausdorff G-space, then $KK^G_0(\mathbb{C},C(M)) \simeq K^0_G(M)$, the Atiyah–Segal equivariant K-theory. In particular, in case M is a point, we obtain that $KK^G_0(\mathbb{C},\mathbb{C}) \simeq R(G)$, the representation ring of G. The following deep theorem allows us to generalize this viewpoint to noncompact groups G.

Theorem 1.7.7 ([15, 16])**.** *Let A,B, and C be G-C^*-algebras. There is an associative, bilinear pairing*

$$KK^G_0(A,B) \times KK^G_0(B,C) \to KK^G_0(A,C)$$

called the Kasparov product.

In particular, this theorem tells us that for a G-C^*-algebra A, the abelian groups $KK_0^G(A,A)$ are in fact rings. Moreover, for a G-C^*-algebra B, the abelian group $KK_0^G(A,B)$ is a left module and $KK_0^G(B,A)$ a right module over the ring $KK_0^G(A,A)$. The case $A = \mathbb{C}$ is of particular interest.

Definition 1.7.8. Let G be a second countable locally compact group. The *Kasparov representation ring* of G is the abelian group $KK_0^G(\mathbb{C},\mathbb{C})$ equipped with the Kasparov product.

We note that for a compact group G, we indeed have an isomorphism of rings $KK_0^G(\mathbb{C},\mathbb{C}) \simeq R(G)$.

For any pair (A,B) of G-C^*-algebras, there are *descent homomorphisms*

$$KK_0^G(A,B) \to KK_0(A \rtimes G, B \rtimes G)$$

compatible with the Kasparov product on both sides. Moreover, there is also an *external product*

$$KK_0^G(A,B) \times KK_0^G(C,D) \to KK_0^G(A \otimes C, B \otimes D),$$

which enjoys good functorial properties as well.

1.7.9 Unbounded Operators

It is sometimes convenient to construct Kasparov modules from unbounded operators. Although the general theory of unbounded operators on Hilbert C^*-modules is quite subtle, we will only need it in a geometric context in which they arise naturally with few technical complications.

Definition 1.7.10. Let (X^0, X^1) be a pair of Hilbert C^*-modules and

$$D : X^0 \supset \operatorname{Dom} D \to X^1,$$

a densely defined operator. We say that D is *regular* if

$$D^* : X^1 \supset \operatorname{Dom} D^* \to X^0,$$

is densely defined and $1 + D^*D$ has a dense range.

For Hilbert space operators, any operator with a densely defined adjoint is regular. Now suppose the modules X^i are G-modules in the sense of Definition 1.7.2. We say the regular operator D is *G-strongly continuous* if G preserves the domains of D and D^*, for $g \in G$ the operators

$$D - (\xi_g^1)^{-1} D \xi_g^0, \quad D^* - (\xi_g^0)^{-1} D^* \xi_g^1,$$

are bounded and the maps

$$g \mapsto D - (\xi_g^1)^{-1} D \xi_g^0, \quad g \mapsto D^* - (\xi_g^0)^{-1} D^* \xi_g^1$$

are continuous for the $*$-strong topology on $\mathrm{Hom}_B^*(X^0, X^1)$.

Definition 1.7.11. Let A and B be G-C^*-algebras. An *unbounded G-Kasparov (A,B)-module* is a triple $(\mathscr{A}, X^0, X^1, D)$, where:

- $\mathscr{A} \subseteq A$ is a dense G-invariant $*$-subalgebra;
- X^i are G-(A,B)-correspondences;
- $D : X^0 \supset \mathrm{Dom}\, D \to X^1$ is a densely defined regular strongly G-continuous operator;

such that:

- $a(1 + D^*D)^{-1} \in \mathbb{K}_B(X)$ for all $a \in \mathscr{A}$.
- For every $a \in \mathscr{A}$, a maps $\mathrm{Dom}\, D$ into $\mathrm{Dom}\, D$, and $[D, a]$ extends to an element of $\mathrm{Hom}_B^*(X^0, X^1)$.

Regularity implies that the operator $(1 + D^*D)^{-1}$ is a bounded adjointable operator. By functional calculus then also $(1+D^*D)^{-1/2}$ is bounded adjointable and the range of $(1 + D^*D)^{-1/2}$ is equal to $\mathrm{Dom}\, D$. For a self-adjoint regular operator, we write $F_D := D(1 + D^2)^{-1/2}$ for its *bounded transform*. The bounded transform captures the relation between bounded and unbounded modules.

Theorem 1.7.12 ([1, Proposition 2.2])**.** *If $(\mathscr{A}, X, D)$ be an unbounded G-Kasparov module, then (X, F_D) is a bounded G-Kasparov module.*

1.7.13 The Dirac-Dual Dirac Method and the γ-Element

Let G be a connected Lie group, $K \subset G$ a maximal compact subgroup, and $X := G/K$ the associated global symmetric space. Assume that X is a G-spin manifold and we denote by $\$ \to X$ its spinor bundle. The bundle $\$$ decomposes as a direct sum $\$ = \$^+ \oplus \$^-$. The Hilbert space $L^2(X, \$)$ then carries a grading by even and odd spinors, and thus decomposes as a direct sum

$$L^2(X, \$) = L^2(X, \$^+) \oplus L^2(X, \$^-).$$

Denote by

$$\not{D} : C_c^\infty(X, \$^+) \to C_c^\infty(X, \$^-),$$

the Dirac operator. Then $(C_c^\infty(X), L^2(X, \$^+), L^2(X, \$^-), D\!\!\!/)$ is a G-equivariant unbounded Kasparov module. Thus, we have an element $\beta \in KK_0^G(C_(X), \mathbb{C})$. Let A be a G-C^*-algebra. Consider the image of β under the composition

$$KK_0^G(C_0(X), \mathbb{C}) \xrightarrow{\tau_A} KK_0^G(C_0(X) \otimes A, A) \xrightarrow{j_G} KK_0(C_0(X) \otimes A \rtimes G, A \rtimes G),$$

where τ_A denotes the external product with $1_A \in KK_0(A, A)$ and j_G is the descent homomorphism. Observe that by Theorem 1.3.23, $C_0(X) \otimes A \rtimes G$ is Morita equivalent to $A \rtimes K$. The *Dirac induction element* is the corresponding element in $KK_0(A \rtimes K, A \rtimes G)$.

Definition 1.7.14. The *Dirac induction homomorphism* is the map

$$\beta_A : K_*(A \rtimes K) \to K_*(A \rtimes G),$$

induced by the Kasparov product with the Dirac induction element.

The *Connes–Kasparov conjecture* is the statement that β_A is an isomorphism for all separable G-C^*-algebras.

Similarly, we consider the Hilbert C^*-$C_0(X)$-module

$$C_0(X, \$) = C_0(X, \$^+) \oplus C_0(X, \$^-)$$

of C_0-sections of the spinor bundle. The space $X = G/K$ has a natural base point $x_0 = [K] \in G/K$ and we denote by $\rho(x)$ the tangent vector to the geodesic connecting x_0 to x of length $d(x_0, x)$. The operator

$$\rho : C_c(X, \$^+) \to C_c(X, \$^-)$$

is then given by Clifford multiplication by $\rho(x)$. Then, $(\mathbb{C}, C_0(X, \$^+), C_0(X, \$^-), \rho)$ is a G-equivariant unbounded Kasparov $(\mathbb{C}, C_0(X))$-module. Thus, we have an element $\alpha \in KK_0^G(\mathbb{C}, C_0(X))$. The Kasparov product $\beta \otimes \alpha$ can be computed.

Theorem 1.7.15 ([16]). $\beta \otimes \alpha = 1 \in KK_0^G(C_0(X), C_0(X))$

In light of this result, the product in the opposite order yields an idempotent in the Kasparov representation ring $KK_0^G(\mathbb{C}, \mathbb{C})$.

Definition 1.7.16. The Kasparov product

$$\gamma := \alpha \otimes_{C_0(X)} \beta \in KK_0^G(\mathbb{C}, \mathbb{C}),$$

is called the *γ-element* for G.

The γ-element restricts to the identity in $KK_0^K(\mathbb{C},\mathbb{C})$. The representation ring decomposes as

$$KK_0^G(\mathbb{C},\mathbb{C}) = \gamma KK_0^G(\mathbb{C},\mathbb{C}) \oplus (1-\gamma)KK_0^G(\mathbb{C},\mathbb{C}),$$

and $\gamma KK_0^G(\mathbb{C},\mathbb{C}) \simeq KK_0^K(\mathbb{C},\mathbb{C})$. The statement that $\gamma = 1$ implies that the Connes–Kasparov conjecture is true. We refer the reader to the excellent survey [11] for an extensive treatment of these topics.

References

[1] S. Baaj and P. Julg. *Théorie bivariante de Kasparov et opérateurs non bornés dans les C^*-modules hilbertiens*. C. R. Acad. Sci. Paris Sér. I Math. 296 (21) (1983), 875–878.

[2] P. Baum, A. Connes, and N. Higson. Classifying space for proper actions and K-theory of group C^*-algebras. In *C^*-algebras: 1943–1993 a fifty year celebration* (ed. R. Doran, San Antonio, TX, 1993), 241–291, Contemporary Mathematics, 167, American Mathematical Society, Providence, RI, 1994.

[3] B. Blackadar. *K-theory for operator algebras*, 2nd ed. Mathematical Sciences Research Institute Publications, 5. Cambridge University Press, Cambridge, 1998.

[4] B. Bekka and P. de la Harpe. *Unitary representations of groups, duals, and characters*. Mathematical Surveys and Monographs, 250. American Mathematical Society, Providence, RI, 2020.

[5] P. Clare. *Hilbert modules associated to parabolically induced representations*. J. Operat. Theor. 69 (2) (2013), 483–509.

[6] P. Clares, T. Crisp, and N. Higson. *Parabolic induction and restriction via C^*-algebras and Hilbert C^*-modules*. Compos. Math. 152 (6) (2016), 1286–1318.

[7] J. Dixmier. *C^*-algebras*. Translated from the French by Francis Jellett. North-Holland Mathematical Library, Vol. 15. North-Holland Publishing Co., Amsterdam-New York-Oxford, 1977.

[8] M. Duflo and C. C. Moore. *On the regular representation of a non-unimodular locally compact group*. J. Func. Anal. 21 (1976), 209–243.

[9] G. B. Folland. *A course in abstract harmonic analysis*. Studies in Advanced Mathematics. CRC Press, Boca Raton, FL, 1995.

[10] M. Frank and D. R. Larson. *Frames in Hilbert C^*-modules and C^*-algebras*. J. Operat. Theor. 48 (2) (2002), 273–314.

[11] M. P. Gomez Aparicio, P. Julg, and A. Valette. The Baum-Connes conjecture: An extended survey. In *Advances in noncommutative geometry – On the occasion of Alain Connes' 70th birthday*, 127–244, Springer, Cham, 2019.

[12] P. Green. *Square-integrable representations and the dual topology*. J. Func. Anat. 35 (1980), 279–294.

[13] R. Howe. *On a notion of rank for unitary representations of the classical groups*. Harmonic Analysis and Group Representations, 223–331, Liguori, Naples, 1982.

[14] G. G. Kasparov. *Hilbert C^*-modules: Theorems of Stinespring and Voiculescu*, J. Operat. Theor. 4 (1980) 133–150.

[15] G. G. Kasparov. *The operator K-functor and extensions of C^*-algebras*. Math. USSR-Izv. 16 (3) (1981), 513–572.
[16] G. G. Kasparov. *Equivariant KK-theory and the Novikov conjecture*. Invent. Math. 91 (1988), 147–201.
[17] E. C. Lance. *Hilbert C^*-modules, a toolkit for operator algebraists*. London Mathematical Society Lecture Note Series, 210. Cambridge University Press, Cambridge, 1995.
[18] J.-S. Li. *Singular unitary representations of classical groups*. Invent. Math. 97 (2) (1989), 237–255.
[19] J.-S. Li. Minimal representations and reductive dual pairs. In *Representation theory of Lie groups* (Park City, UT, 1998), 293–340, IAS/Park City Mathematics Series, 8, American Mathematical Society, Providence, RI, 2000.
[20] V. M. Manuilov and E. V. Troitsky. *Hilbert C^*-modules*. Translated from the 2001 Russian original by the authors. Translations of Mathematical Monographs, 226. American Mathematical Society, Providence, RI, 2005.
[21] B. Mesland and M. H. Şengün. Stable range local theta correspondence via C^*-correspondences. Preprint.
[22] D. Miličić. *Topological representation of the group C^*-algebra of* SL*(2,R)*. Glasnik Mat. Ser. III 6 (26) (1971), 231–246.
[23] I. Raeburn and D. P. Williams. *Morita equivalence and continuous trace C^*-algebras*. Mathematical Surveys and Monographs, 60. American Mathematical Society, Providence, RI, 1998.
[24] F. Pierrot. *Induction parabolique et K-théorie de C^*-algèbres maximales*. C. R. Acad. Sci. Paris Sér. I Math. 332 (9) (2001), 805–808.
[25] M. A. Rieffel. *Induced representations of C^*-algebras*. Adv. Math. 13 (1974), 176–257.
[26] M. A. Rieffel. *Strong Morita equivalence of certain transformation group C^*-algebras*. Math. Ann. 222 (1) (1976), 7–22.
[27] M. A. Rieffel. Morita equivalence for operator algebras. In *Operator algebras and applications*, Part 1 (Kingston, ON, 1980), pp. 285–298, Proceedings of Symposia in Pure Mathematics, 38, American Mathematical Society, Providence, RI, 1982.
[28] M. A. Rieffel. Applications of strong Morita equivalence to transformation group C^*-algebras. In *Operator algebras and applications*, Part 1 (Kingston, ON, 1980), 299–310, Proceedings of Symposia in Pure Mathematics, 38, American Mathematical Society, Providence, RI, 1982.
[29] M. Rørdam, F. Larsen, and N. Lautsten. *An introduction to K-theory for C^*-algebras*. London Mathematical Society Student Texts, 49. Cambridge University Press, Cambridge, 2000.
[30] D. Soudry. *Explicit Howe duality in the stable range*. J. Reine Angew. Math. 396 (1989), 70–86.
[31] A. Valette. *Minimal projections, integrable representations and property (T)*. Arch. Math. (Basel) 43 (5) (1984), 397–406.
[32] S. P. Wang. *On isolated points in the dual spaces of locally compact groups*. Math. Ann. 218 (1975), 19–34.

2

Tempered Representations of Semisimple Lie Groups

Peter Hochs

2.1 Introduction

This chapter is about the representation theory needed to describe the reduced C^*-algebra of a real semisimple Lie group and its K-theory. Our first goal is Theorem 2.5.24: a classification of almost all irreducible tempered representations of a connected, semisimple, linear Lie group. We then apply this to find a description of reduced group C^*-algebras in Section 2.6.

To reach this goal, we discuss some general concepts that are useful throughout Lie theory and representation theory: basics on (linear) Lie groups and Lie algebras; Cartan subgroups and roots; basics on representations; normalized induction of representations; (cuspidal) parabolic subgroups; intertwining operators and R-groups.

This chapter is intended as an overview of the theory of tempered representations of semisimple Lie groups. It contains some short proofs where they are illustrative, but we give references for most of the results that we use.

2.1.1 Notation

We write $\mathbb{N}$ for the set of natural numbers, starting at 1.

For $n \in \mathbb{N}$, let $M_n(\mathbb{R})$ and $M_n(\mathbb{C})$ be the vector spaces of real and complex $n \times n$ matrices, respectively. We write $\mathrm{GL}(n,\mathbb{R})$ and $\mathrm{GL}(n,\mathbb{C})$ for the groups of invertible matrices in these spaces. We denote the transpose of a matrix $A \in M_n(\mathbb{C})$ by A^t, and its conjugate transpose by A^*. The identity matrix in $M_n(\mathbb{R}) \subset M_n(\mathbb{C})$ is denoted by I or by I_n.

If V is a finite-dimensional vector space over $\mathbb{R}$ or $\mathbb{C}$, then $\mathrm{End}(V)$ will denote the algebra of linear endomorphisms of V. The notation V^* always means the space of real linear maps from V to $\mathbb{R}$, even if V may be viewed as a vector space over $\mathbb{C}$. So iV^* is the space of real linear maps from $V \to i\mathbb{R}$, and $V^* \otimes \mathbb{C}$ is the space of real linear maps from V to $\mathbb{C}$.

If V and W are two normed vector spaces, then $\mathscr{B}(V,W)$ denotes the space of bounded linear maps from V to W, and we write $\mathscr{B}(V) := \mathscr{B}(V,V)$ for the algebra of bounded linear endomorphisms of V.

The identity element of a group is denoted by e.

2.2 Lie Groups and Lie Algebras

2.2.1 Real Semisimple Lie Groups

Definition 2.2.2. A *linear Lie group* is a closed subgroup of $\mathrm{GL}(n,\mathbb{C})$, for some n. A linear Lie group $G < \mathrm{GL}(n,\mathbb{C})$ is *semisimple* if it is closed under the conjugate transpose operation, and has a finite center.

We will often consider *connected*, linear, semisimple Lie groups, although we will also need disconnected groups in some places. The class of connected, semisimple, linear Lie groups is large enough to contain interesting examples and small enough to allow for an explicit and concise treatment.

Much of the theory in this chapter generalizes to larger classes of groups, however. For example, connectedness may sometimes be weakened to a condition like the one at the start of Subsection 2.4.13. Groups satisfying the conditions of Definition 2.2.2 but with an infinite center are *reductive*, and still have many of the properties that we are going to discuss. And one can go beyond linear Lie groups to a more general class of groups, that don't admit embeddings into a group of matrices. See also Harish-Chandra's set of conditions on G at the start of Section 3 of [14]. For the sake of concreteness, we will mainly restrict ourselves to connected, semisimple, linear Lie groups.

Example 2.2.3. The *orthogonal group* is

$$\mathrm{O}(n) := \{g \in \mathrm{GL}(n,\mathbb{R}); g^t g = I\}.$$

The *special orthogonal group* is

$$\mathrm{SO}(n) := \{g \in \mathrm{O}(n); \det(g) = 1\}.$$

The groups $\mathrm{O}(n)$ and $\mathrm{SO}(n)$ have finite centers if $n \geq 3$; then they are linear, semisimple Lie groups. The group $\mathrm{O}(n)$ has two connected components, of matrices with determinants ± 1, and $\mathrm{SO}(n)$ is connected.

The *unitary group* is

$$\mathrm{U}(n) := \{g \in \mathrm{GL}(n,\mathbb{C}); g^* g = I\}.$$

The *special unitary group* is

$$\mathrm{SU}(n) := \{g \in \mathrm{U}(n); \det(g) = 1\}.$$

The group $\mathrm{U}(n)$ has an infinite center $\{zI; z \in \mathrm{U}(1)\}$, so it is not semisimple. The center of $\mathrm{SU}(n)$ is cyclic of order n, and $\mathrm{SU}(n)$ is a connected, linear, semisimple Lie group.

Consider the matrix

$$J_n := \begin{pmatrix} 0 & I_n \\ -I_n & 0 \end{pmatrix} \in M_{2n}(\mathbb{R}). \tag{2.2.3.1}$$

The group

$$\mathrm{Sp}(n) := \{g \in \mathrm{U}(2n); g^t J_n g = J_n\}$$

is connected and semisimple.

The groups $\mathrm{O}(n)$, $\mathrm{SO}(n)$, $\mathrm{U}(n)$, $\mathrm{SU}(n)$, and $\mathrm{Sp}(n)$ are all *compact*.

Example 2.2.4. The group $\mathrm{GL}(n,\mathbb{C})$ itself has an infinite center $\{zI; z \in \mathrm{GL}(1,\mathbb{C})\}$, so it is not semisimple.

The *complex special linear group* is

$$\mathrm{SL}(n,\mathbb{C}) := \{g \in \mathrm{GL}(n,\mathbb{C}); \det(g) = 1\}.$$

This is a connected, linear, semisimple Lie group. Its center is cyclic of order n.

The *complex special orthogonal group*

$$\mathrm{SO}(n,\mathbb{C}) := \{g \in \mathrm{GL}(n,\mathbb{C}); g^t g = I, \det(g) = 1\}$$

is also a connected, linear, semisimple Lie group if $n \geq 3$.

The *complex symplectic group*

$$\mathrm{Sp}(n,\mathbb{C}) := \{g \in \mathrm{SL}(2n,\mathbb{C}); g^t J_n g = J_n\},$$

with J_n as in (2.2.3.1), is connected and semisimple.

The groups in this example are *complex*; see Definition 2.2.15. They are not compact.

Example 2.2.5. The group $\mathrm{GL}(n,\mathbb{R})$ has an infinite center $\{zI; z \in \mathrm{GL}(1,\mathbb{R})\}$, so it is not semisimple.

The *real special linear group* is

$$\mathrm{SL}(n,\mathbb{R}) = \mathrm{SL}(n,\mathbb{C}) \cap \mathrm{GL}(n,\mathbb{R}).$$

This is a connected, linear, semisimple Lie group. It is noncompact if $n \geq 2$. There is also the *quaternionic special linear group* $\mathrm{SL}(n,\mathbb{H})$ of $n \times n$ matrices over the quaternions with determinant 1. It is also connected and semisimple.

(Matrices over the quaternions may be viewed as matrices over the complex numbers of twice the size; see the end of Section I.8 in [21].)

Let $p, q \in \mathbb{N}$. Let $I_{p,q}$ be the diagonal matrix

$$I_{p,q} = \begin{pmatrix} I_p & 0 \\ 0 & -I_q \end{pmatrix} \in M_{p+q}(\mathbb{R}).$$

The group

$$\mathrm{U}(p,q) = \{g \in \mathrm{GL}(p+q, \mathbb{C}); g^* I_{p,q} g = I_{p,q}\}$$

has an infinite center, but its subgroup

$$\mathrm{SU}(p,q) = \mathrm{U}(p,q) \cap \mathrm{SL}(p+q, \mathbb{C}) \tag{2.2.5.1}$$

has a finite center. The latter is a connected, semismple, linear Lie group.

The group

$$\mathrm{O}(p,q) = \mathrm{U}(p,q) \cap \mathrm{GL}(p+q, \mathbb{R})$$

is disconnected and noncompact. Its subgroup

$$\mathrm{SO}(p,q) = \mathrm{O}(p,q) \cap \mathrm{SL}(p+q, \mathbb{R}) \tag{2.2.5.2}$$

has two connected components; for example, in the group $\mathrm{SO}(1,1)$, the matrices $\pm I_2$ lie in different connected components. The groups $\mathrm{O}(p,q)$ and $\mathrm{SO}(p,q)$ are noncompact, disconnected, linear, semisimple Lie groups. The connected component $\mathrm{SO}_0(p,q) < \mathrm{SO}(p,q)$ containing the identity is a noncompact, connected, linear semisimple Lie group.

The analogue of (2.2.5.1) and (2.2.5.2) over the quaternions is

$$\mathrm{Sp}(p,q) := \{g \in \mathrm{SL}(p+q, \mathbb{H}); g^t I_{p,q} g = I_{p,q}\}.$$

It is a connected, semisimple, linear Lie group. (It is linear because an $n \times n$ matrix over $\mathbb{H}$ can be identified with a $2n \times 2n$ matrix over $\mathbb{C}$.)

The *symplectic group*

$$\mathrm{Sp}(n, \mathbb{R}) = \{g \in \mathrm{SL}(2n, \mathbb{R}); g^t J_n g = J_n\}$$

is a connected, linear, semisimple Lie group. Here J_n is as in (2.2.3.1).

Let

$$\tilde{J}_n := \begin{pmatrix} 0 & I_n \\ I_n & 0 \end{pmatrix}.$$

The group

$$\mathrm{SO}^*(2n) := \{g \in \mathrm{SU}(n,n); g^t \tilde{J}_{n,n} g = \tilde{J}_{n,n}\}$$

is also a connected, semisimple, linear Lie group.

The groups in this example are neither compact nor complex.

See Section I.14 of [21] for more information about the groups in Examples 2.2.3 to 2.2.5, including proofs of connectedness.

Example 2.2.6. The groups of upper-triangular matrices in $\mathrm{GL}(n,\mathbb{R})$ and $\mathrm{GL}(n,\mathbb{C})$ with ones on their diagonals are not closed under conjugate transpose. So they are not semisimple.

The fundamental group of $\mathrm{SL}(2,\mathbb{R})$ is $\mathbb{Z}$. The universal cover of $\mathrm{SL}(2,\mathbb{R})$ is not a subgroup of $\mathrm{GL}(n,\mathbb{C})$ for any n, so it is not linear. Its center is infinite.

Definition 2.2.7. If $G < \mathrm{GL}(n,\mathbb{C})$ is a linear, semisimple Lie group, then its *maximal compact subgroup* is $K = G \cap \mathrm{U}(n)$.

The group K in this definition is maximal compact in the sense that it is compact and not contained in a strictly larger compact subgroup. In general, one calls a subgroup with the properties a maximal compact subgroup, but we will use the specific subgroup of Definition 2.2.7 and call it *the* maximal compact subgroup. Any two maximal compact subgroups are conjugate, so it is no limitation to make this choice.

Example 2.2.8. The maximal compact subgroup of $\mathrm{SL}(n,\mathbb{C})$ is $\mathrm{SU}(n)$; the maximal compact subgroup of both $\mathrm{SO}(n,\mathbb{C})$ and $\mathrm{SL}(n,\mathbb{R})$ is $\mathrm{SO}(n)$. The maximal compact subgroup of $\mathrm{Sp}(n,\mathbb{R})$ is $\mathrm{U}(n) < \mathrm{GL}(n,\mathbb{C}) < \mathrm{GL}(2n,\mathbb{R})$.

For the groups $\mathrm{SU}(p,q)$ and $\mathrm{SO}(p,q)$, the maximal compact subgroups are $\mathrm{S}(\mathrm{U}(p) \times \mathrm{U}(q))$ and $\mathrm{S}(\mathrm{O}(p) \times \mathrm{O}(q))$, where we write $\mathrm{S}(G) = G \cap \mathrm{SL}(n,\mathbb{C})$ for $G < \mathrm{GL}(n,\mathbb{C})$. For the connected group $\mathrm{SO}_0(p,q)$, the maximal compact subgroup is $\mathrm{SO}(p) \times \mathrm{SO}(q)$.

Let G be a linear, semisimple Lie group. We write $C_c(G)$ for the space of complex-valued, continuous functions on G with compact supports. Like any locally compact topological group, G admits a *left Haar measure*: a measure dg such that for all $f \in C_c(G)$ and $g' \in G$,

$$\int_G f(g'g)\,dg = \int_G f(g)\,dg. \tag{2.2.8.1}$$

Any two left Haar measures differ by a constant factor. We fix the left Haar measures for all groups we use. This allows us to define the Hilbert space $L^2(G)$ of square-integrable, complex-valued functions on G (modulo equality almost everywhere).

The group G is *unimodular*, meaning that a left Haar measure is also a right Haar measure: for all $f \in C_c(G)$ and $g' \in G$,

$$\int_G f(gg')\,dg = \int_G f(g)\,dg. \tag{2.2.8.2}$$

2.2.9 Lie Algebras

The *exponential map* $\exp\colon M_n(\mathbb{C}) \to \mathrm{GL}(n,\mathbb{C})$ is given by

$$\exp(X) = \sum_{j=0}^{\infty} \frac{1}{j!} X^j,$$

for $X \in M_n(\mathbb{C})$.

Let $G < \mathrm{GL}(n,\mathbb{C})$ be a linear Lie group.

Definition 2.2.10. The *Lie algebra* of G is

$$\mathfrak{g} = \{X \in M_n(\mathbb{C}); \exp(tX) \in G \text{ for all } t \in \mathbb{R}\}.$$

Example 2.2.11. The unitary and orthogonal groups $\mathrm{U}(n)$ and $\mathrm{O}(n)$ have Lie algebras

$$\begin{aligned} \mathfrak{u}(n) &= \{X \in M_n(\mathbb{C}); X^* + X = 0\} \text{ and} \\ \mathfrak{o}(n) &= \{X \in M_n(\mathbb{R}); X^t + X = 0\}, \end{aligned} \tag{2.2.11.1}$$

respectively.

The special linear groups $\mathrm{SL}(n,F)$, for $F = \mathbb{R}$ or $F = \mathbb{C}$, have Lie algebras

$$\mathfrak{sl}(n,F) = \{X \in M_n(F); \mathrm{tr}(X) = 0\}. \tag{2.2.11.2}$$

This follows from the equality $\det(\exp(X)) = e^{\mathrm{tr}(X)}$ for all $X \in M_n(\mathbb{C})$. It follows from (2.2.11.1) and 2.2.11.2 that the groups $\mathrm{SU}(n)$ and $\mathrm{SO}(n)$ have Lie algebras

$$\begin{aligned} \mathfrak{su}(n) &= \mathfrak{u}(n) \cap \mathfrak{sl}(n,\mathbb{C}) = \{X \in M_n(\mathbb{C}); X^* + X = 0, \mathrm{tr}(X) = 0\}; \\ \mathfrak{so}(n) &= \mathfrak{o}(n) \cap \mathfrak{sl}(n,\mathbb{R}) = \{X \in M_n(\mathbb{R}); X^t + X = 0\}. \end{aligned} \tag{2.2.11.3}$$

The groups $\mathrm{SO}_0(p,q)$, $\mathrm{SU}(p,q)$ and $\mathrm{Sp}(n)$ have Lie algebras

$$\begin{aligned} \mathfrak{su}(p,q) &= \{X \in M_n(\mathbb{C}); X^* I_{p,q} + I_{p,q} X = 0, \mathrm{tr}(X) = 0\}; \\ \mathfrak{so}(p,q) &= \{X \in M_n(\mathbb{R}); X^t I_{p,q} + I_{p,q} X = 0\}; \\ \mathfrak{sp}(n) &= \{X \in M_{2n}(\mathbb{R}); X^t J_n + J_n X = 0\}. \end{aligned} \tag{2.2.11.4}$$

Lemma 2.2.12. *The Lie algebra of a linear Lie group is closed under addition, scalar multiplication, conjugation by elements in G, and the commutator bracket given by*

$$[X,Y] = XY - YX$$

for $X,Y \in M_n(\mathbb{C})$.

Proof. Closedness under scalar multiplication and conjugation is immediate from the definition. Closedness under addition follows from

$$\exp(X+Y) = \lim_{n\to\infty} (\exp(Y/n)\exp(Y/n))^n$$

for all $X,Y \in M_n(\mathbb{C})$. Closedness under the commutator bracket follows from closedness under conjugation by elements of G, and

$$[X,Y] = \left.\frac{d}{dt}\right|_{t=0} \exp(tX)Y\exp(-tX) = \lim_{t\to 0} \frac{\exp(tX)Y\exp(-tX) - Y}{t}.$$

□

Definition 2.2.13. For a linear Lie group G, the *adjoint actions* $\mathrm{Ad}\colon G \to \mathrm{GL}(\mathfrak{g})$ and $\mathrm{ad}\colon \mathfrak{g} \to \mathrm{End}(\mathfrak{g})$ are given by

$$\begin{aligned} \mathrm{Ad}(g)X &= gXg^{-1}; \\ \mathrm{ad}(X)Y &= [X,Y], \end{aligned} \tag{2.2.13.1}$$

for $g \in G$ and $X,Y \in \mathfrak{g}$.

When we talk about "a Lie algebra," we mean the Lie algebra of some linear Lie group. Or equivalently, about a real-linear subspace of $M_n(\mathbb{C})$ closed under the commutator bracket. A semisimple Lie algebra will mean the Lie algebra of a linear semisimple Lie group. This is the same thing as a real linear subspace of $M_n(\mathbb{C})$ closed under the commutator bracket, with a trivial center, which is closed under the conjugate transpose operation. (See Propositions 1.97 and 1.98 in [21].)

Definition 2.2.14. A Lie algebra $\mathfrak{g}$ is *commutative* if $[X,Y] = 0$ for all $X,Y \in \mathfrak{g}$. It is *complex* if it is a complex linear subspace of $M_n(\mathbb{C})$. A *subalgebra* of $\mathfrak{g}$ is a real linear subspace closed under the commutator bracket.

Definition 2.2.15. A linear Lie group G is *complex* if its Lie algebra is complex.

Definition 2.2.16. The *centralizer* of a subalgebra $\mathfrak{h} \subset \mathfrak{g}$ in a subgroup $H' < G$ is

$$Z_{H'}(\mathfrak{h}) := \{h \in H'; \mathrm{Ad}(h)X = X \text{ for all } X \in \mathfrak{h}\}.$$

The centralizer of $\mathfrak{h}$ in a subalgebra $\mathfrak{h}' \subset \mathfrak{g}$ is

$$Z_{\mathfrak{h}'}(\mathfrak{h}) := \{Y \in \mathfrak{h}'; [Y,X] = 0 \text{ for all } X \in \mathfrak{h}\}.$$

Definition 2.2.17. Let $\mathfrak{n}$ be a Lie algebra. Define $\mathfrak{n}_0 := \mathfrak{n}$, and for $j \in \mathbb{N}$, set

$$\mathfrak{n}_j := [\mathfrak{n}, \mathfrak{n}_{j-1}] = \{[X,Y]; X \in \mathfrak{n}, Y \in \mathfrak{n}_{j-1}\}.$$

Then $\mathfrak{n}$ is *nilpotent* if $\mathfrak{n}_j = \{0\}$ for some $j \in \mathbb{N}$.

A connected, linear Lie group is nilpotent if its Lie algebra is.

Example 2.2.18. The Lie algebra of upper-triangular matrices in $M_n(F)$, for $F = \mathbb{R}$ or $F = \mathbb{C}$,

$$\left\{ \begin{pmatrix} 0 & x_{12} & \dots & x_{1n} \\ 0 & 0 & \dots & x_{2n} \\ \vdots & & \ddots & \vdots \\ 0 & \dots & 0 & 0 \end{pmatrix} ; x_{jk} \in F \right\}.$$

is nilpotent. It is the Lie algebra of

$$\left\{ \begin{pmatrix} 1 & x_{12} & \dots & x_{1n} \\ 0 & 1 & \dots & x_{2n} \\ \vdots & & \ddots & \vdots \\ 0 & \dots & 0 & 1 \end{pmatrix} ; x_{jk} \in F \right\} < \mathrm{SL}(n,F),$$

which therefore is a nilpotent, linear Lie group.

2.2.19 Cartan Subgroups

The Lie bracket on a Lie algebra $\mathfrak{g}$, and hence the adjoint representation ad, extends complex-linearly to its complexification $\mathfrak{g} \otimes \mathbb{C}$.

Definition 2.2.20. Let G be a linear semisimple Lie group.

- A *Cartan subalgebra* of $\mathfrak{g}$ is a maximal commutative subalgebra $\mathfrak{h} \subset \mathfrak{g}$ (i.e., $Z_{\mathfrak{g}}(\mathfrak{h}) = \mathfrak{h}$) such that the set of operators

$$\{\mathrm{ad}(X); X \in \mathfrak{h}\}$$

  on $\mathfrak{g} \otimes \mathbb{C}$ diagonalizes simultaneously.
- A *Cartan subgroup* of G is the centralizer in G of a Cartan subalgebra.

Lemma 2.2.21. *A linear semisimple Lie group has finitely many conjugacy classes of Cartan subgroups. A compact or complex linear semisimple Lie group has only one conjugacy class of Cartan subgroups. In general, all Cartan subgroups of a given linear semisimple Lie group have the same dimension.*

For the compact case, see Proposition 4.30, Theorem 4.34, and Corollary 4.35 in [21]. For the complex case, and the equality of dimensions, see Theorem 2.15 in [21]. The general case is Proposition 6.64 in [21].

Remark 2.2.22. *Because compact and complex linear semisimple Lie groups only have one conjugacy class of Cartan subgroups, the representation theory for such groups is less complicated than for general linear semisimple Lie groups.*

Definition 2.2.23. The *rank* of a linear semisimple Lie group is the (real) dimension of a Cartan subgroup.

Example 2.2.24. All Cartan subgroups of $\mathrm{SU}(n)$ and $\mathrm{SL}(n,\mathbb{C})$ are conjugate to the subgroups of diagonal matrices in these groups. So $\mathrm{rank}(\mathrm{SU}(n)) = n-1$ and $\mathrm{rank}(\mathrm{SL}(n,\mathbb{C})) = 2n-2$.

The Cartan subgroups of $\mathrm{SO}(2n)$ are conjugate to the group $\mathrm{SO}(2)^n$, embedded into $\mathrm{SO}(2n)$ as 2×2 blocks along the diagonal. The Cartan subgroups of $\mathrm{SO}(2n+1)$ of are conjugate to the group $\mathrm{SO}(2)^n$, embedded into $\mathrm{SO}(2n)$ as 2×2 blocks along the diagonal with a 1 in the bottom right entry. The ranks of these groups are n.

The group $\mathrm{SL}(2,\mathbb{R})$ has two conjugacy classes of Cartan subgroups. One of these is represented by $\mathrm{SO}(2)$, and the other by the subgroup of diagonal matrices. So $\mathrm{rank}(\mathrm{SL}(2,\mathbb{R})) = 1$.

Let $K < G$ be the maximal compact subgroup. Write

$$\mathfrak{s} := \{X \in \mathfrak{g}; X^* = X\}.$$

Then we have the *Cartan decomposition*

$$\mathfrak{g} = \mathfrak{k} \oplus \mathfrak{s}. \tag{2.2.24.1}$$

Definition 2.2.25. A Cartan subalgebra $\mathfrak{h} \subset \mathfrak{g}$ is *θ-stable* if

$$\mathfrak{h} = (\mathfrak{h} \cap \mathfrak{k}) \oplus (\mathfrak{h} \cap \mathfrak{s}).$$

Then the corresponding Cartan subgroup is also called θ-stable.

Theorem 2.2.26 (Matsuki, 1979). *Every Cartan subgroup is conjugate to a θ-stable Cartan subgroup. Two θ-stable Cartan subgroups are conjugate in G if and only if they are conjugate by an element of K.*

See Proposition 6.59 in [21] for the first statement. For the second statement, see Theorem 2 in [26]; see also Proposition 6.18 [4] or Proposition 6.15 in [1].

2.2.27 Roots

Definition 2.2.28. Let $\mathfrak{g}$ be a semisimple Lie algebra, and $\mathfrak{h} \subset \mathfrak{g}$ a Cartan subalgebra.

- For $\alpha \in \mathfrak{h}^* \otimes \mathbb{C}$, write
$$\mathfrak{g}_\alpha^{\mathbb{C}} = \{X \in \mathfrak{g} \otimes \mathbb{C}; [Y, X] = \langle \alpha, Y \rangle X \text{ for all } Y \in \mathfrak{h}\}. \qquad (2.2.28.1)$$
- A nonzero element $\alpha \in \mathfrak{h}^* \otimes \mathbb{C}$ is a *root* of $(\mathfrak{g}, \mathfrak{h})$ if $\mathfrak{g}_\alpha^{\mathbb{C}} \neq 0$.
- The *root system* of $(\mathfrak{g}, \mathfrak{h})$ is the set $R(\mathfrak{g}, \mathfrak{h})$ of roots of $(\mathfrak{g}, \mathfrak{h})$.

Lemma 2.2.29. *The complexification of a semisimple Lie algebra $\mathfrak{g}$, with a Cartan subalgebra $\mathfrak{h}$ decomposes as*
$$\mathfrak{g} \otimes \mathbb{C} = (\mathfrak{h} \otimes \mathbb{C}) \oplus \bigoplus_{\alpha \in R(\mathfrak{g}, \mathfrak{h})} \mathfrak{g}_\alpha^{\mathbb{C}}. \qquad (2.2.29.1)$$
The root spaces $\mathfrak{g}_\alpha^{\mathbb{C}}$ are one-dimensional over $\mathbb{C}$.

See Propositions 2.5 and 2.21 in [21].

For a root $\alpha \in R(\mathfrak{g}, \mathfrak{h})$, the space (2.2.28.1) is the corresponding *root space*. And (2.2.29.1) is the *root space decomposition* of $\mathfrak{g} \otimes \mathbb{C}$ with respect to the Cartan subalgebra $\mathfrak{h}$.

Example 2.2.30. Consider the compact Cartan subgroup $T = \mathrm{SO}(2)$ of $G = \mathrm{SL}(2, \mathbb{R})$. Its Lie algebra $\mathfrak{t}$ is spanned by
$$Y_1 := \begin{pmatrix} 0 & -1 \\ 1 & 0 \end{pmatrix}.$$
Now $R(\mathfrak{g}, \mathfrak{t}) = \{\pm\alpha\}$, where $\langle \alpha, Y_1 \rangle = 2i$. We have
$$\begin{aligned} \mathfrak{g}_\alpha^{\mathbb{C}} &= \mathbb{C}\begin{pmatrix} 1 & -i \\ -i & -1 \end{pmatrix}; \\ \mathfrak{g}_{-\alpha}^{\mathbb{C}} &= \mathbb{C}\begin{pmatrix} 1 & i \\ i & -1 \end{pmatrix}. \end{aligned} \qquad (2.2.30.1)$$
The corresponding root space decomposition is
$$\mathfrak{g} \otimes \mathbb{C} = \mathfrak{sl}(2, \mathbb{C}) = \mathbb{C}Y_1 \oplus \mathbb{C}\begin{pmatrix} 1 & -i \\ -i & -1 \end{pmatrix} \oplus \mathbb{C}\begin{pmatrix} 1 & i \\ i & -1 \end{pmatrix}.$$

Example 2.2.31. A noncompact Cartan subgroup of $G = \mathrm{SL}(2, \mathbb{R})$ is
$$H := \left\{ \begin{pmatrix} x & 0 \\ 0 & x^{-1} \end{pmatrix}; x \neq 0 \right\}.$$

Its Lie algebra is the Cartan subalgebra $\mathfrak{h} \subset \mathfrak{g}$ spanned by

$$Y_2 := \begin{pmatrix} 1 & 0 \\ 0 & -1 \end{pmatrix}. \tag{2.2.31.1}$$

Now $R(\mathfrak{g}, \mathfrak{h}) = \{\pm\beta\}$, where $\langle \beta, Y_2 \rangle = 2$. We have

$$\begin{aligned} \mathfrak{g}_\beta^{\mathbb{C}} &= \mathbb{C}\begin{pmatrix} 0 & 1 \\ 0 & 0 \end{pmatrix}; \\ \mathfrak{g}_{-\beta}^{\mathbb{C}} &= \mathbb{C}\begin{pmatrix} 0 & 0 \\ 1 & 0 \end{pmatrix}. \end{aligned} \tag{2.2.31.2}$$

The root space decomposition is

$$\mathfrak{sl}(2, \mathbb{C}) = \mathbb{C}Y_2 \oplus \mathfrak{g}_\beta^{\mathbb{C}} \oplus \mathfrak{g}_{-\beta}^{\mathbb{C}}.$$

Lemma 2.2.32. *Let $\mathfrak{g}$ be a semisimple Lie algebra, and $\mathfrak{h} \subset \mathfrak{g}$ a Cartan subalgebra. If $\alpha \in R(\mathfrak{g}, \mathfrak{h})$, then $-\alpha \in R(\mathfrak{g}, \mathfrak{t})$.*

See Proposition 2.17(c) in [21].

Because of Lemmas 2.2.29 and 2.2.32, the root system of a semisimple Lie algebra with respect to some Cartan subalgebra is a finite subset of $\mathfrak{h}^* \otimes \mathbb{C}$ that is invariant under reflection in the origin.

Definition 2.2.33. Let $\mathfrak{g}$ be a semisimple Lie algebra, and $\mathfrak{h} \subset \mathfrak{g}$ a Cartan subalgebra. A *positive root system* for $(\mathfrak{g}, \mathfrak{h})$ is a subset $R^+(\mathfrak{g}, \mathfrak{h}) \subset R(\mathfrak{g}, \mathfrak{h})$ that lies on one side of a real hyperplane through the origin in $\mathfrak{h}^* \otimes \mathbb{C}$ that is disjoint from $R(\mathfrak{g}, \mathfrak{h})$.

If $R^+(\mathfrak{g}, \mathfrak{h}) \subset R(\mathfrak{g}, \mathfrak{h})$ is a positive root system, then $R^+(\mathfrak{g}, \mathfrak{h})$ and $-R^+(\mathfrak{g}, \mathfrak{h})$ are disjoint. And Lemma 2.2.32 implies that

$$R(\mathfrak{g}, \mathfrak{h}) = R^+(\mathfrak{g}, \mathfrak{h}) \cup -R^+(\mathfrak{g}, \mathfrak{h}).$$

Definition 2.2.34. The *Weyl group* associated to a linear semisimple Lie group G and a Cartan subgroup $H < G$ is

$$W(G, H) = N_G(H)/H,$$

where $N_G(H) < G$ is the normalizer of H in G.

In the setting of Definition 2.2.34, there is an action by $W(G, H)$ on $\mathfrak{h}^* \otimes \mathbb{C}$, given by

$$\langle w \cdot \xi, X \rangle = \langle \xi, \mathrm{Ad}(g)^{-1} X \rangle \tag{2.2.34.1}$$

for $w \in W$, represented by $g \in N_G(H)$, $\xi \in \mathfrak{h}^* \otimes \mathbb{C}$ and $X \in \mathfrak{h}$.

Lemma 2.2.35. *Let G be a linear semisimple Lie group, and $H < G$ a Cartan subgroup. The Weyl group $W(G,H)$ is finite, and the action 2.2.34.1 preserves $R(G,H) \subset \mathfrak{h}^* \otimes \mathbb{C}$.*

Proof. If $\alpha \in R(G,H)$, $g \in N_G(H)$, $X \in \mathfrak{g}_\alpha^\mathbb{C}$ and $Y \in \mathfrak{h}$, then

$$[Y, \mathrm{Ad}(g)X] = \langle gH \cdot \alpha, \mathrm{Ad}(g)X\rangle,$$

so $\mathrm{Ad}(g)X \in \mathfrak{g}^\mathbb{C}_{gH\cdot\alpha}$. We find that $W(G,H)$ preserves $R(G,H)$.

Because $R(G,H)$ spans $\mathfrak{h}^* \otimes \mathbb{C}$ over $\mathbb{C}$ (Proposition 2.17(e) in [21]), an element $gH \in W(G,H)$ acts trivially on $R(G,H)$ if and only if $\mathrm{Ad}(g)$ acts trivially on $\mathfrak{h}$. This means that $g \in H$. So the homomorphism from $W(G,H)$ to the symmetric group of $R(G,H)$ defined by the action (2.2.34.1) is injective, hence $W(G,H)$ is finite. □

2.2.36 Restricted Roots and Nilpotent Subalgebras

Definition 2.2.37. Let $\mathfrak{g}$ be a semisimple Lie algebra, and $\mathfrak{a} \subset \mathfrak{g}$ a commutative subalgebra of self-adjoint matrices. For $\alpha \in \mathfrak{a}^*$, write

$$\mathfrak{g}_\alpha = \{X \in \mathfrak{g}; [Y,X] = \langle\alpha, Y\rangle X \text{ for all } Y \in \mathfrak{a}\}. \tag{2.2.37.1}$$

A nonzero element $\alpha \in \mathfrak{a}^*$ is a *restricted root* of $(\mathfrak{g},\mathfrak{a})$ if $\mathfrak{g}_\alpha \neq 0$. Then (2.2.37.1) is the *restricted root space* corresponding to α. The *restricted root system* of $(\mathfrak{g},\mathfrak{a})$ is the set $\Sigma(\mathfrak{g},\mathfrak{a})$ of restricted roots of $(\mathfrak{g},\mathfrak{a})$.

Differences between roots and restricted roots are that there is no complexification in Definition 2.2.37, and the subalgebra $\mathfrak{a} \subset \mathfrak{g}$ need not be a Cartan subalgebra. A consequence is that the restricted root spaces (2.2.37.1) may have dimensions higher than one.

Consider the inner product on $\mathfrak{g}$ given by

$$(X,Y) = \mathrm{Re}(\mathrm{tr}(XY^*)), \tag{2.2.37.2}$$

for $X,Y \in \mathfrak{g}$. For $X \in \mathfrak{g}$, the adjoint of $\mathrm{ad}(X)$ with respect to this inner product is $\mathrm{ad}(X^*)$. So

$$\{\mathrm{ad}(X); X \in \mathfrak{a}\}$$

is a commuting set of self-adjoint operators on $\mathfrak{g}$. We therefore have the *restricted root space decomposition*

$$\mathfrak{g} = Z_\mathfrak{g}(\mathfrak{a}) \oplus \bigoplus_{\alpha\in\Sigma(\mathfrak{g},\mathfrak{a})} \mathfrak{g}_\alpha. \tag{2.2.37.3}$$

Lemma 2.2.38. *Let $\mathfrak{g}$ be a semisimple Lie algebra, and $\mathfrak{a} \subset \mathfrak{g}$ a commutative subalgebra of self-adjoint matrices. For all $\alpha, \beta \in \Sigma(\mathfrak{g}, \mathfrak{a})$, we have*

- $[\mathfrak{g}_\alpha, \mathfrak{g}_\alpha] \subset \mathfrak{g}_{\alpha+\beta}$;
- $-\alpha \in \Sigma(\mathfrak{g}, \mathfrak{a})$.

See Proposition 6.40(b),(c) in [21].

A positive restricted root system is a subset $\Sigma^+(\mathfrak{g}, \mathfrak{a}) \subset \Sigma(\mathfrak{g}, \mathfrak{a})$ lying on one side of a hyperplane in $\mathfrak{a}^*$ not intersecting $\Sigma(\mathfrak{g}, \mathfrak{a})$. By the second point in Lemma 2.2.38,

$$\Sigma(\mathfrak{g}, \mathfrak{a}) = \Sigma^+(\mathfrak{g}, \mathfrak{a}) \cup -\Sigma^+(\mathfrak{g}, \mathfrak{a}).$$

By the first point in Lemma 2.2.38, the subspace

$$\bigoplus_{\alpha \in \Sigma^+(\mathfrak{g}, \mathfrak{a})} \mathfrak{g}_\alpha \subset \mathfrak{g} \tag{2.2.38.1}$$

is a subalgebra. By finiteness of $\Sigma(\mathfrak{g}, \mathfrak{a})$, this subalgebra is nilpotent.

Example 2.2.39. Let $\mathfrak{g} = \mathfrak{sl}(n, F)$, with $F = \mathbb{R}$ or $\mathbb{C}$. Consider the subalgebra $\mathfrak{a} \subset \mathfrak{g}$ of diagonal matrices with real entries. For $j = 1, \ldots, n$, let $\xi_j \in \mathfrak{a}^*$ be the functional mapping a matrix in $\mathfrak{a}$ to its jth diagonal element. Then

$$\Sigma(\mathfrak{g}, \mathfrak{a}) = \{\xi_j - \xi_k; j, k = 1, \ldots, n, j \neq k\}.$$

For different $j, k \in \{1, \ldots, n\}$, the corresponding restricted root space consists of all matrices whose entries other than the one at position (j, k) are zero.

A choice of positive restricted roots is

$$\Sigma^+(\mathfrak{g}, \mathfrak{a}) = \{\xi_j - \xi_k; j, k = 1, \ldots, n, j < k\}.$$

The corresponding nilpotent subalgebra (2.2.38.1) is the algebra $\mathfrak{n}$ of upper-triangular matrices, from Example 2.2.18.

2.3 Tempered Representations

2.3.1 Representations

Let G be a linear Lie group. (Most of the contents of this section apply to more general topological groups.)

Definition 2.3.2. Let V be a complete topological vector space over $\mathbb{C}$, and $\mathrm{GL}(V)$ the group of all linear maps from V to itself. A *(continuous) representation* of G in V is a group homomorphism

$$\pi \colon G \to \mathrm{GL}(V)$$

such that the map $G \times V \to V$, mapping $(g, v) \in G \times V$ to $\pi(g)v$, is continuous.

If V is a Hilbert space, and $\pi(G) \subset \mathrm{U}(V)$, then such a representation is *unitary*.

If V is finite-dimensional, then π is a *finite-dimensional* representation.

All representations will tacitly be assumed to be continuous. Sometimes we talk about a "representation π" of G, without explicitly mentioning the space V. Then we write V_π for this representation space. We will only consider representations in Hilbert spaces, though they may not always be unitary.

Example 2.3.3. The inclusion map of G into $\mathrm{GL}(n,\mathbb{C})$ is a finite-dimensional representation of G. It is unitary if and only if G is compact.

Example 2.3.4. The *trivial representation* 1_G of G in $\mathbb{C}$, given by $1_G(g) = 1$ for all $g \in G$, is unitary.

Let $L^2(G)$ be the Hilbert space of square-integrable, complex-valued functions on G with respect to the Haar measure dg.

Example 2.3.5. The *left regular representation* L and the *right regular representation* R of G in $L^2(G)$ are given by

$$\begin{aligned}(L(g)f)(g') &= f(g^{-1}g');\\ (R(g)f)(g') &= f(g'g),\end{aligned} \qquad (2.3.5.1)$$

for $g,g' \in G$ and $f \in L^2(G)$. The left regular representation is unitary by (2.2.8.1). If G is unimodular, then so is the right regular representation, by (2.2.8.2). These two representations commute and combine into a representation $L \times R$ of $G \times G$ in $L^2(G)$.

Definition 2.3.6. A representation π of G in V is *irreducible* if the only closed linear subspaces of V that are invariant under $\pi(G)$ are $\{0\}$ and V.

Definition 2.3.7. For $j = 1,2$, let π_j be a representation of G in a vector space V_j.

- A linear operator $T\colon V_1 \to V_2$ is *intertwining* if for all $g \in G$,

$$\pi_2(g) \circ T = T \circ \pi_1(g).$$

- The representations π_1 and π_2 are *equivalent* if there is an intertwining isomorphism of topological vector spaces $T\colon V_1 \to V_2$.
- The *unitary dual* of G is the set $\hat{G}$ of equivalence classes of irreducible unitary representations of G.

We often choose representatives of equivalence classes in $\hat{G}$ without mentioning this explicitly.

Example 2.3.8. Let $K < G$ be a compact subgroup, different from $\{I\}$ and G. The closed subspace

$$L^2(G)^{R(K)} = \{f \in L^2(G); f(gk) = f(g) \text{ for all } g \in G \text{ and } k \in K\} \subset L^2(G)$$

is invariant under $L(G)$. This space is neither $\{0\}$ nor $L^2(G)$, so the left regular representation of G is *not* irreducible.

Example 2.3.9. Let $G = \mathbb{R}^n$; then no subgroup K as in Example 2.3.8 exists. Let $f \in L^2(\mathbb{R}^n)$ be such that its Fourier transform is supported in a compact set $C \subset \mathbb{R}^n$. Then the Fourier transform of $(L \times R)(g,g')f$ is supported in the same set C, for all $g,g' \in \mathbb{R}^n$. So all functions in the closure of the span of these functions $(L \times R)(g,g')f$ are supported in C. Hence, this closure is not all of $L^2(\mathbb{R}^n)$, and invariant under $L \times R$. That representation is therefore not irreducible.

If π_j is a representation of G in a vector space V_j, for $j = 1,2$, then the direct sum representation $\pi_1 \oplus \pi_2$ of G in $V_1 \oplus V_2$ is defined by

$$(\pi_1 \oplus \pi_2)(g)(v_1,v_2) = (\pi(g)v_1,\pi(g)v_2),$$

for $g \in G$ and $v_j \in V_j$. If $(\pi_j)_{j=1}^{\infty}$ is a countable set of unitary representations π_j of G in Hilbert spaces V_j, then we can analogously define the unitary representation $\bigoplus_j \pi_j$ of G in the Hilbert space direct sum $\bigoplus_j V_j$. It is generally not true that a unitary representation decomposes as a direct sum of irreducible unitary representations; see Examples 2.3.8 and 2.3.9. But for compact groups, such decompositions do exist (see Lemma 2.3.13), and much of representation theory is simpler than for noncompact groups.

2.3.10 Compact Groups

In this subsection, we suppose that K is a compact, linear Lie group.

Lemma 2.3.11. *Every irreducible unitary representation of K is finite-dimensional.*

See Theorem 12(c) in [20], or [24, 30] for more elementary proofs.

Example 2.3.12. Let π be any representation of K in a Hilbert space V. Define the inner product $(-,-)_K$ on V by

$$(v,w)_K := \int_K (\pi(k)v,\pi(k)w)_V \, dk.$$

Then π is a unitary representation with respect to the inner product $(-,-)_K$.

Lemma 2.3.13. *Every unitary representation of K decomposes as a Hilbert space direct sum of irreducible representations.*

See Theorem 1.12(d) in [20].

Let G be a connected, semisimple, linear Lie group, and suppose that K is the maximal compact subgroup of G. A unitary representation π of G restricts to a unitary representation $\pi|_K$ of the maximal compact subgroup $K < G$. By Lemma 2.3.13, that restriction decomposes as a direct sum of irreducible representations. A priori, a given irreducible representation of K may appear infinitely often in this decomposition. That is not the case if π is irreducible.

Theorem 2.3.14. *If π is a unitary irreducible representation of a connected, semisimple, linear Lie group G, with maximal compact subgroup K, then*

$$\pi|_K = \bigoplus_{\tau \in \hat{K}} m_\tau^\pi \tau, \tag{2.3.14.1}$$

where $0 \leq m_\tau^\pi \leq \dim(V_\tau)$, and $m_\tau^\pi \tau$ is the direct sum of m_τ^π copies of τ.

See Theorem 8.1 in [20].

Definition 2.3.15. In the setting of Theorem 2.3.14, the representations $\tau \in \hat{K}$ for which $m_\tau^\pi > 0$ are the *K-types* of π.

Definition 2.3.16. Let T be an abelian, linear Lie group, and $\exp_T : \mathfrak{t} \to T$ its exponential map. An element $\xi \in i\mathfrak{t}^*$ is *analytically integral* if for all $X \in \ker(\exp_T)$,

$$\langle \xi, X \rangle \in 2\pi i \mathbb{Z}.$$

Lemma 2.3.17. *For a connected, abelian, linear Lie group T, there is a bijection from the set $\Lambda(T)$ of analytically integral elements of $i\mathfrak{t}^*$ onto $\hat{T}$, given by $\xi \mapsto e^\xi$, where for $X \in \mathfrak{t}$,*

$$e^\xi(\exp(X)) = e^{\langle \xi, X \rangle} \in \mathrm{U}(\mathbb{C}). \tag{2.3.17.1}$$

Proof. It follows from Schur's lemma (Lemma 2.5.2 later) that all irreducible representations of an abelian group are one-dimensional. The rest of the argument is elementary. □

Lemma 2.3.18. *Let π be a finite-dimensional, continuous representation of K in V. For $v \in V$, the map from $\mathfrak{g}$ to V, given by*

$$X \mapsto \pi(\exp(X))v$$

for $X \in \mathfrak{g}$, is smooth.

See Corollary 3.16 in [20].

In the setting of Lemma 2.3.18, we define

$$\pi(X)v := \left.\frac{d}{dt}\right|_{t=0} \pi(\exp(tX))v$$

for $X \in \mathfrak{g}$ and $v \in V$. This defines a linear map $\pi\colon \mathfrak{g} \to \mathrm{End}(V)$. We denote its complex-linear extension by

$$\pi^{\mathbb{C}}\colon \mathfrak{g} \otimes \mathbb{C} \to \mathrm{End}(V).$$

For non-abelian, compact, connected, linear Lie groups K, irreducible unitary representations are classified by their highest weights. Let $T < K$ be a Cartan subgroup, and choose a positive root system for $(\mathfrak{g}, \mathfrak{t})$.

Definition 2.3.19. Let π be an irreducible unitary representation of K in V. An element $\lambda \in i\mathfrak{t}^*$ is a *highest weight* of π if

- there is a nonzero $v \in V$ such that for all $X \in \mathfrak{t}$,

$$\pi(X)v = \langle \lambda, X \rangle v$$

- for all $\alpha \in R^+(\mathfrak{g}, \mathfrak{t})$ and all $X \in \mathfrak{g}^{\mathbb{C}}_{\alpha}$,

$$\pi^{\mathbb{C}}(X)v = 0.$$

Fix an inner product $(-,-)$ on $i\mathfrak{t}^*$ invariant under the action (2.2.34.1) of the Weyl group $W(K,T)$. This can be constructed as in Example 2.3.12.

Definition 2.3.20. An element $\xi \in i\mathfrak{t}^*$ is *algebraically integral* if $2\frac{(\xi,\alpha)}{(\alpha,\alpha)}$ is an integer for all $\alpha \in R(\mathfrak{k}, \mathfrak{t})$. It is *dominant* with respect to the chosen positive root system $R^+(\mathfrak{k}, \mathfrak{t})$ if $(\xi, \alpha) \geq 0$ for all $\alpha \in R^+(\mathfrak{k}, \mathfrak{t})$.

An analytically integral element of $i\mathfrak{t}^*$ is algebraically integral; see Proposition 4.59 in [21].

Theorem 2.3.21 (Theorem of the highest weight; Weyl, 1925). *Let K be a compact, connected, linear Lie group. Fix a Cartan subgroup and a positive root system. For all dominant, algebraically integral $\lambda \in i\mathfrak{t}^*$, there is an irreducible representation $\pi_\lambda \in \hat{K}$ with the highest weight λ. No two of these are equivalent, and all irreducible representations of K occur in this way.*

The original result is in [36]. This is based on a result by Cartan for representations of Lie algebras [7]. See also Theorem 4.28 in [20] or Theorem 5.5 in [21].

2.3.22 The Plancherel Theorem

Let (X,μ) be a measure space. Suppose that, for every $x \in X$, a Hilbert space V_x is given. We write

$$\int_X^{\oplus} V_x \, d\mu(x)$$

for the Hilbert space of maps (modulo equality almost everywhere)

$$\varphi \colon X \to \coprod_{x \in X} V_x$$

such that $\varphi(x) \in V_x$ for all $x \in X$, such that the map $x \mapsto \|\varphi(x)\|_{V_x}$ is measurable, and

$$\int_X \|\varphi(x)\|_{V_x}^2 \, d\mu(x)$$

converges. The inner product on this Hilbert space is given by

$$(\varphi,\psi) = \int_X (\varphi(x),\psi(x))_{V_x} \, d\mu(x),$$

for $\varphi,\psi \in \int_X^{\oplus} V_x \, d\mu(x)$.

For a Hilbert space V, we write V^* for its continuous dual space; this is again a Hilbert space. If W is another Hilbert space, then $V \otimes W$ is the Hilbert space tensor product of these spaces. In particular, $V \otimes V^*$ is the Hilbert space of Hilbert–Schmidt operators on V.

Let G be a separable, unimodular, locally compact topological group, with a (left and right) Haar measure dg. If $\pi \in \hat{G}$, and $f \in C_c(G)$, then we write

$$\hat{f}(\pi) = \pi(f) = \int_G f(g)\pi(g) \, dg.$$

Because the operator norm of the integrand is at most $|f|$, the integral converges to a bounded operator on V_π. In this way, we obtain the *Fourier transform* of f: the map

$$\hat{f} \colon \hat{G} \to \coprod_{\pi \in \hat{G}} \mathcal{B}(V_\pi)$$

such that $\hat{f}(\pi) \in \mathcal{B}(V_\pi)$ for all $\pi \in \hat{G}$.

The space $L^2(G)$ carries the unitary representation $L \times R$ of $G \times G$ from Example 2.3.5. Given a measure μ on $\hat{G}$, we have the natural unitary representation of $G \times G$ in the Hilbert space

$$\int_{\hat{G}} V_\pi \otimes V_\pi^* \, d\mu(\pi),$$

given by

$$((g,g')\cdot\varphi)(\pi) = \pi(g)\circ\varphi(\pi)\circ\pi(g'^{-1}),$$

for $g,g' \in G$, $\varphi \in \int_{\hat{G}} V_\pi \otimes V_\pi^*\, d\mu(\pi)$ and $\pi \in \hat{G}$. Here, we view $\varphi(\pi) \in V_\pi \otimes V_\pi^*$ as a Hilbert–Schmidt operator on V_π.

The general Plancherel theorem applies to a much more general class of groups than the connected, semisimple, linear Lie groups we consider in most of this chapter. See 13.9.4 in [11] for the definition of type I groups, and Theorem 7 in [12], or [34], for the result that (linear) semisimple Lie groups are of type I.

Theorem 2.3.23 (Abstract Plancherel theorem). *Suppose that G is a unimodular, locally compact group of type I. For all $\pi \in \hat{G}$ and $f \in L^1(G)\cap L^2(G)$, the operator $\hat{f}(\pi)$ lies in the space $V_\pi \otimes V_\pi^*$ of Hilbert–Schmidt operators. And there is a unique measure μ on $\hat{G}$ such that the Fourier transform extends to a $G\times G$ equivariant unitary isomorphism*

$$L^2(G) \to \int_{\hat{G}} V_\pi \otimes V_\pi^*\, d\mu(\pi). \qquad (2.3.23.1)$$

See Theorems 18.8.1 and 18.8.2 in [11], or [27, 28, 33].

For connected, semisimple, linear Lie groups, Harish-Chandra's Plancherel theorem [14–16] is a much more explicit version of Theorem 2.3.23. It includes descriptions of the measure μ and its support in $\hat{G}$.

Definition 2.3.24. The *Fell topology* on $\hat{G}$ is defined by the property that a sequence $(\pi_j)_{j=1}^\infty$ in $\hat{G}$ converges to $\pi \in \hat{G}$ precisely if for all $v \in V_\pi$, there are $v_j \in V_{\pi_j}$ such that for every compact subset $X \subset G$,

$$\lim_{j\to\infty} \max_{g\in X}|(v,\pi(g)v)_{V_\pi} - (v_j,\pi_j(g)v_j)_{V_{\pi_j}}| = 0.$$

Definition 2.3.25. Let G be a unimodular, locally compact group of type I. The measure μ in Theorem 2.3.23 is the *Plancherel measure* of G. A unitary irreducible representation of G is *tempered* if it is in the support of the Plancherel measure.

In other words, $\pi \in \hat{G}$ is tempered if and only if every neighborhood of π in the Fell topology has positive Plancherel measure.

Now let G be a connected, semisimple, linear Lie group. Let $K < G$ be the maximal compact subgroup.

Definition 2.3.26. Let π be a representation of G in a vector space V. A vector $v \in V$ is *K-finite* if its K-orbit $\pi(K)v$ spans a finite-dimensional subspace of V.

If V is a Hilbert space and $\pi|_K$ is unitary, then Lemma 2.3.13 implies that the space of K-finite vectors is dense in V.

Theorem 2.3.27. *Suppose that G is a connected, semisimple, linear Lie group. Then a representation $\pi \in \hat{G}$ is tempered if and only if the function*

$$g \mapsto (v, \pi(g)w)_V \tag{2.3.27.1}$$

lies in $L^{2+\varepsilon}(G)$ for all K-finite vectors $v, w \in V_\pi$ and $\varepsilon > 0$.

See [10].

A large part of Harish-Chandra's work is to compute the Plancherel measure for semisimple Lie groups. This includes a classification of almost all tempered representations. Our main goal in this chapter is to review that classification; see Theorem 2.5.24.

Example 2.3.28. If G is *compact*, then the Peter–Weyl theorem (Theorem 4.20 in [21]) states that the Plancherel measure is the counting measure on $\hat{G}$. So the Fourier transform defines a $G \times G$-equivariant unitary isomorphism

$$L^2(G) \cong \bigoplus_{\pi \in \hat{G}} V_\pi \otimes V_\pi^*.$$

Example 2.3.29. If $G = \mathbb{R}^n$, then

$$\hat{G} = \{e^{i\xi}; \xi \in (\mathbb{R}^n)^*\}$$

where for $g \in G$,

$$e^{i\xi}(g) = e^{i\langle \xi, g\rangle} \in \mathrm{U}(\mathbb{C}).$$

Now the Plancherel measure is the Lebesgue measure on $(\mathbb{R}^n)^* \cong \mathbb{R}^n$, and the Fourier transform is the classical Fourier transform.

In this example, the functions (2.3.27.1) never lie in $L^p(\mathbb{R}^n)$ for any finite $p > 0$. This illustrates that the condition in Theorem 2.3.27 that G is semisimple is necessary.

Example 2.3.30. Suppose that $G = \mathrm{SL}(2, \mathbb{R})$. Its irreducible, unitary representations are

- the *trivial representation* (Example 2.3.4);
- the *complementary series* $\{C_x; 0 < x < 1\}$;
- the *spherical principal series* $\{P_\nu^+; \nu \geq 0\}$;
- the *non spherical principal series* $\{P_\nu^-; \nu > 0\}$;
- the *discrete series* $\{D_n^\pm; n \in \mathbb{N}\}$;
- the two *limits of discrete series* representations $D_0^\pm$.

The trivial representation and the complementary series are not tempered; the other representations on this list are. Every discrete series representation $D_n^{\pm}$ has Plancherel measure $n/2\pi$. The limits of discrete series representations have Plancherel measure zero. On the spherical principal series, the Plancherel measure is the Lebesgue measure on $[0,\infty)$ times the weight function

$$P_\nu^+ \mapsto \frac{1}{8\pi}\tanh(\pi\nu/2).$$

On the nonspherical principal series, the Plancherel measure is the Lebesgue measure on $(0,\infty)$ times the weight function

$$P_\nu^- \mapsto \frac{1}{8\pi}\coth(\pi\nu/2).$$

We will discuss discrete series representations in Subsection 2.3.31 and principal series representations in Subsection 2.4.15. See Subsection 2.3.38 for an explicit description of the discrete series of $\mathrm{SL}(2,\mathbb{R})$, and Example 2.4.33 for the principal series. We finish the classification of the tempered representations of $\mathrm{SL}(2,\mathbb{R})$ below Example 2.5.37.

2.3.31 The Discrete Series

Let G be a connected, semisimple, linear Lie group.

Proposition 2.3.32. *Let $\pi \in \hat{G}$. The following are equivalent:*

1. *for all $v, w \in V_\pi$, the function $g \mapsto (v, \pi(g)w)_{V_\pi}$ lies in $L^2(G)$;*
2. *the Plancherel measure of $\{\pi\}$ is positive.*

See Proposition 9.6 in [20] or Theorem 18.8.5 in [11] for a more general statement. The idea is that if the first condition holds, then there is a $G \times G$-equivariant embedding $V_\pi \otimes V_\pi^* \hookrightarrow L^2(G)$, given by

$$v \otimes \xi \mapsto (g \mapsto \langle \xi, \pi(g)v\rangle).$$

So $V_\pi \otimes V_\pi^*$ occurs as a direct summand on the right hand side of (2.3.23.1).

Definition 2.3.33. A unitary irreducible representation of G belongs to the *discrete series* if it satisfies the equivalent conditions of Proposition 2.3.32.

Example 2.3.34. Every unitary irreducible representation of a compact group belongs to the discrete series.

Theorem 2.3.35 (Harish-Chandra)**.** *The group G has discrete series representations if and only if it has a compact Cartan subgroup.*

Table 2.1 *Harish–Chandra's criterion* rank(G) = rank(K) *for the existence of discrete series representations, for the (noncompact, connected) classical groups*

Group G	Max. cpt. $K < G$	rank(G)	rank(K)	Discrete series?
$\mathrm{SL}(n,\mathbb{C})$	$\mathrm{SU}(n)$	$2n-2$	$n-1$	no
$\mathrm{SO}(n,\mathbb{C})$	$\mathrm{SO}(n)$	$2\lfloor\frac{n}{2}\rfloor$	$\lfloor\frac{n}{2}\rfloor$	no
$\mathrm{Sp}(n,\mathbb{C})$	$\mathrm{Sp}(n)$	$2n$	n	no
$\mathrm{SL}(n,\mathbb{R})$	$\mathrm{SO}(n)$	$n-1$	$\lfloor\frac{n}{2}\rfloor$	iff $n=2$
$\mathrm{SL}(n,\mathbb{H})$	$\mathrm{Sp}(n)$	$2n-1$	n	no
$\mathrm{SU}(p,q)$	$\mathrm{S}(\mathrm{U}(p)\times\mathrm{U}(q))$	$p+q-1$	$p+q-1$	yes
$\mathrm{SO}_0(p,q)$	$\mathrm{SO}(p)\times\mathrm{SO}(q)$	$\lfloor\frac{p+q}{2}\rfloor$	$\lfloor\frac{p}{2}\rfloor+\lfloor\frac{q}{2}\rfloor$	iff pq even
$\mathrm{Sp}(n,\mathbb{R})$	$\mathrm{U}(n)$	n	n	yes
$\mathrm{Sp}(p,q)$	$\mathrm{Sp}(p)\times\mathrm{Sp}(q)$	$p+q$	$p+q$	yes
$\mathrm{SO}^*(2n)$	$\mathrm{U}(n)$	n	n	yes

See Theorem 12.20 in [20] or Theorem 13 in [13]. If $K < G$ is the maximal compact subgroup, then a compact Cartan subgroup of G is conjugate to a compact Cartan subgroup of K. So G has a compact Cartan subgroup if and only if rank(G) = rank(K). See Table 2.1 for some examples.

Example 2.3.36. Suppose that G is a complex, semisimple, linear Lie group. Then every Cartan subalgebra $\mathfrak{h} \subset \mathfrak{g}$ is a complex-linear subspace of $M_n(\mathbb{C})$. So $\mathfrak{h}$ is not contained in $\mathfrak{u}(n)$, and H is not compact. Therefore, G does not have discrete series representations.

Suppose that G has a compact Cartan subgroup T that lies inside K. Then $R(\mathfrak{g},\mathfrak{t}) \subset i\mathfrak{t}^*$. Choose an inner product $(-,-)$ on $i\mathfrak{t}^*$ that is invariant under the action (2.2.34.1) by the finite group $W(G,T)$. An element $\lambda \in i\mathfrak{t}^*$ is *regular* if $(\lambda,\alpha) \neq 0$ for all $\alpha \in R(\mathfrak{g},\mathfrak{t})$. If $\lambda \in i\mathfrak{t}^*$ is regular, then it defines positive root systems

$$\begin{aligned} R^+_\lambda(\mathfrak{g},\mathfrak{t}) &:= \{\alpha \in R(\mathfrak{g},\mathfrak{t}); (\lambda,\alpha) > 0\}; \\ R^+_\lambda(\mathfrak{k},\mathfrak{t}) &:= \{\alpha \in R(\mathfrak{k},\mathfrak{t}); (\lambda,\alpha) > 0\}. \end{aligned} \tag{2.3.36.1}$$

For such a λ, we write

$$\begin{aligned} \rho^G_\lambda &:= \frac{1}{2}\sum_{\alpha\in R^+_\lambda(\mathfrak{g},\mathfrak{t})} \alpha; \\ \rho^K_\lambda &:= \frac{1}{2}\sum_{\alpha\in R^+_\lambda(\mathfrak{k},\mathfrak{t})} \alpha. \end{aligned} \tag{2.3.36.2}$$

Theorem 2.3.37 (Harish-Chandra, 1966)**.** *Suppose that G has a compact Cartan subgroup T.*

(*a*) *For every regular* $\lambda \in i\mathfrak{t}^*$ *such that* $\lambda + \rho_\lambda^G$ *is analytically integral, there is a discrete series representation* π_λ *of G for which the highest weights (with respect to* $R_\lambda^+(\mathfrak{k},\mathfrak{t})$*) of all K-types are of the form*

$$\lambda + \rho_\lambda^G - 2\rho_\lambda^K + \sum_{\alpha \in R_\lambda^+(\mathfrak{g},\mathfrak{t})} n_\alpha \alpha,$$

for $n_\alpha \geq 0$.

(*b*) *Two discrete series representations* π_λ *and* $\pi_{\lambda'}$ *as in part* (*a*) *are equivalent if and only if there is a* $w \in W(K,T)$ *such that* $\lambda' = w \cdot \lambda$.

(*c*) *Every discrete series representation of G occurs as in part* (*a*).

See Theorem 9.20 in [20] or Theorem 16 in [13] for the original formulation.

2.3.38 Example: The Discrete Series of SL(2, ℝ)

Consider the group $G = \mathrm{SL}(2,\mathbb{R})$ and the compact Cartan subgroup $T = \mathrm{SO}(2)$. We use the notation from Example 2.2.30. If $\lambda = t\alpha/2 \in i\mathfrak{t}^*$, for $t \in \mathbb{R}$, then

$$\langle \lambda, Y_1 \rangle = it.$$

Now $(\lambda,\alpha) \neq 0$ if and only if $t \neq 0$. And

$$\rho_\lambda^G = \begin{cases} \alpha/2 & \text{if } t > 0; \\ -\alpha/2 & \text{if } t < 0. \end{cases}$$

Now

$$\exp(sY_1) = \begin{pmatrix} \cos s & -\sin s \\ \sin s & \cos s \end{pmatrix},$$

so $\ker(\exp_T) = 2\pi\mathbb{Z}Y_1$.

Let $\lambda = t\alpha/2$ with $t > 0$. Then $\lambda + \rho^\lambda = (t+1)\alpha/2$. This element is analytically integral if and only if

$$2\pi i(t+1) = \langle \lambda + \rho^\lambda, 2\pi Y_1 \rangle \in 2\pi i\mathbb{Z},$$

so $t = n \in \mathbb{N} = \{1,2,3,\ldots\}$. We have the corresponding *holomorphic discrete series* representation D_n^+ of $\mathrm{SL}(2,\mathbb{R})$.

If $\lambda = t\alpha/2$ with $t < 0$, then $\lambda + \rho^\lambda = (t-1)\alpha/2$. This element is analytically integral if and only if $-t = n \in \mathbb{N}$. We have the corresponding *antiholomorphic discrete series* representation D_n^-.

Now $K = T$, so $W(K,T)$ is trivial. Therefore, none of the discrete series representations D_n^+ and D_n^- are equivalent.

Let $n \in \mathbb{N}$. Consider the inner product

$$(f_1, f_2)_n := \int_{\mathbb{H}} f_1(z)\overline{f_2(z)}\, \mathrm{Im}(z)^{n-1}\, dz$$

for functions on the upper half plane $\mathbb{H} = \{\mathrm{Im}(z) > 0\}$. Consider the Hilbert space

$$V_n^+ := \{f : \mathbb{H} \to \mathbb{C}; \text{holomorphic}, (f, f)_n < \infty\}.$$

Then D_n^+ is the representation of $\mathrm{SL}(2,\mathbb{R})$ in V_n^+ given by

$$(D_n^+(g)f)(z) = (-bz + d)^{-n-1} f\left(\frac{az - c}{-bz + d}\right).$$

for $g = \begin{pmatrix} a & b \\ c & d \end{pmatrix} \in \mathrm{SL}(2,\mathbb{R})$.

Let $V_n^- = \overline{V_n^+}$ be the Hilbert space defined as V_n^+ but with antiholomorphic functions. The representation D_n^- is the representation in V_n^- given by the same expression as D_n^+.

2.3.39 Other Classes of Irreducible Representations

The discrete series, and the tempered representations, of a semisimple, linear Lie group G, are relevant because of their roles in the Plancherel decomposition of $L^2(G)$. But they are also used in the classification of larger classes of representations.

A not necessarily unitary representation of G in a Hilbert space is *admissible* if its restriction to a maximal compact subgroup is unitary, and all irreducible representations of K occur finitely many times in the decomposition of this restriction as in Lemma 2.3.13. Non admissible representations can behave very wildly, and are not studied as much as admissible representations. And Theorem 2.3.14 states that the important class of unitary irreducible representations lies inside the class of admissible irreducible representations. So there are inclusions

$$\begin{aligned} \text{discrete series} &\subset \text{tempered representations} \\ &\subset \text{unitary irreducible representations} \\ &\subset \text{admissible irreducible representations} \\ &\subset \text{irreducible representations.} \end{aligned}$$

For compact groups, all inclusions are equalities.

We will see in Theorem 2.5.24 that almost all tempered representations can be classified in terms of the discrete series of certain subgroups. These subgroups may be disconnected, so one needs a generalization of Theorem 2.3.37; see Theorem 2.4.14. For a complete classification of tempered representations, one also needs the *limits of discrete series* [22, 23].

In a similar way, admissible irreducible representations can be classified in terms of tempered representations of similar subgroups. This is the *Langlands classification*, see Theorem 8.54 in [20] or [5, 25]. This means that admissible irreducible representations have been classified, and the classification of unitary irreducible representations has been reduced to the problem of determining which admissible irreducible representations can be made unitary. That is a very deep problem that has not been solved yet. The ATLAS software package [2, 3] contains an algorithm that determines if a given individual admissible representation can be made unitary.

2.4 Parabolic Induction

Induction is a way to construct representations of a group from representations of a subgroup. We are particularly interested in *normalized induction*, which is a version of induction for which the induced representation of a unitary representation is again unitary. We will later apply this to the *cuspidal parabolic subgroups* of semisimple, linear Lie groups that are introduced in Section 2.4.15.

2.4.1 Modular Functions

Consider a locally compact topological group G. Let dg be a left Haar measure on G, as in (2.2.8.1). It need not be right invariant as in (2.2.8.2) in general.

Definition 2.4.2. The *modular function* for dg is the function $\Delta_G : G \to (0, \infty)$ such that for all $f \in C_c(G)$ and $g' \in G$,

$$\int_G f(gg')\, dg = \Delta_G(g')^{-1} \int_G f(g)\, dg.$$

So G is unimodular if and only if Δ_G is constant 1. If G is a linear Lie group, then for all $g \in G$,

$$\Delta_G(g) = |\det(\mathrm{Ad}(g))|^{-1}. \qquad (2.4.2.1)$$

(See Proposition 8.27 in [21].) It follows directly from the definition that Δ_G is a group homomorphism with respect to multiplication in $(0, \infty)$.

Let $H < G$ be a closed subgroup. Fix a left Haar measure dh on H, and let Δ_H be the corresponding modular function. Consider the group homomorphism

$$\chi := \frac{\Delta_G^{1/2}|_H}{\Delta_H^{1/2}} : H \to (0,\infty). \tag{2.4.2.2}$$

Lemma 2.4.3. *Let f be a locally integrable function on G such that for all $g \in G$ and $h \in H$,*

$$f(gh) = \chi(h)^{-2} f(g).$$

Let $\psi \in C_c(G)$ be such that for all $g \in G$,

$$\int_H \psi(gh)\, dh = 0.$$

Then

$$\int_G \psi(g) f(g)\, dg = 0.$$

The proof is a short computation, see Proposition 1.13 in [17].

Lemma 2.4.4. *Suppose that G is unimodular, and that there is a closed subgroup $H' < G$ such that $H' \cap H$ is compact, and $H'H = G$. Then there is a left Haar measure dh' on H' such that for all $f \in C_c(G)$,*

$$\int_G f(g)\, dg = \int_{H'} \int_H f(h'h) \Delta_H(h)^{-1}\, dh\, dh'.$$

Proof. Consider the action by $H' \times H$ on itself given by

$$(h_1', h_1) \cdot (h_2', h_2) = (h_1' h_2', h_2 h_1^{-1}),$$

for $h_1, h_2 \in H$ and $h_1', h_2' \in H'$. Consider the action by $H' \cap H$ on $H' \times H$ given by

$$x \cdot (h', h) = (h' x^{-1}, xh),$$

for $x \in H' \cap H$, $h \in H$, and $h' \in H'$. The two actions commute, and the multiplication map descends to a homeomorphism

$$X := (H' \times H)/(H' \cap H) \cong G.$$

Because $H' \cap H$ is compact, we have $C_c(X) \cong C_c(H' \times H)^{H' \cap H}$, where the superscript $H' \cap H$ denotes $(H' \cap H)$-invariant functions. Let dh' initially be any left Haar measure on H'. Then the two linear functionals I_1 and I_2 on $C_c(X) \cong C_c(G)$ mapping a function f to

$$\begin{aligned} I_1(f) &= \int_G f(g)\,dg; \\ I_2(f) &= \int_{H'}\int_H f([h',h])\Delta_H(h)^{-1}\,dh\,dh' \end{aligned} \tag{2.4.4.1}$$

are both positive and $(H' \times H)$-invariant. (For I_1, we use the unimodularity of G here; for I_2, we use the right invariance of the measure $\Delta_H^{-1}\,dh$ on H.) By the Riesz representation theorem, these functionals are given by integration with respect to $(H' \times H)$-invariant Radon measures on X. Because the action by $H \times H'$ on X is transitive, these measures differ by a constant. Rescaling dh' by the inverse of this constant, we can ensure that $I_1 = I_2$. □

2.4.5 Normalized Induction

Let π be a unitary representation of H in a Hilbert space V. We write $\mathbb{C}_\chi$ for the complex numbers, equipped with the representation (2.4.2.2) of H. Let $C(G, V \otimes \mathbb{C}_\chi)^H$ be the space of continuous functions $F\colon G \to V$ whose supports have compact images in G/H, and such that for all $g \in G$ and $h \in H$,

$$F(gh) = \chi(h)^{-1}\pi(h)^{-1}F(g). \tag{2.4.5.1}$$

To define an inner product on $C(G, V \otimes \mathbb{C}_\chi)^H$, the idea is to use an L^2-type inner product on G/H. Explicitly, let $F_1, F_2 \in C(G, V \otimes \mathbb{C}_\chi)^H$. Let $\psi \in C_c(G)$ be a nonnegative function such that

$$\int_H \psi(gh)\,dh = 1$$

for all $g \in G$ such that $gH \in G/H$ lies in the image of the support of F_1 or F_2. (Such a function always exists; see Proposition 1.9 in [17].) Then we set

$$(F_1, F_2) := \int_G \psi(g)(F_1(g), F_2(g))_V\,dg. \tag{2.4.5.2}$$

Lemma 2.4.6. *The expression* (2.4.5.2) *is an inner product on* $C(G, V \otimes \mathbb{C}_\chi)^H$, *independent of the choice of* ψ.

Proof. Nondegeneracy follows from the transformation property (2.4.5.1) of functions in $C(G, V \otimes \mathbb{C}_\chi)^H$. Independence of ψ follows from Lemma 2.4.3; here we use unitarity of π and the factor $\mathbb{C}_\chi$ in the definition of $C(G, V \otimes \mathbb{C}_\chi)^H$. □

Remark 2.4.7. *Suppose that G is a linear Lie group. Then*

$$(C(G)\otimes\mathbb{C}_{\chi^2})^H := \{f \in C(G); f(gh) = \chi(h)^{-2}f(g) \text{ for all } g \in G \text{ and } h \in H\}$$

is the space of continuous densities *on G/H. For $F_1, F_2 \in C(G, V \otimes \mathbb{C}_\chi)^H$, the pointwise inner product function*

$$g \mapsto (F_1(g), F_2(g))_V$$

lies in $(C(G) \otimes \mathbb{C}_{\chi^2})^H$. So it is a compactly supported density on G/H and has a natural integral over G/H; this integral equals (2.4.5.2) *for a suitable normalization of dg.*

Definition 2.4.8. The Hilbert space $\mathrm{Ind}_H^G(V)$ is the completion of $C(G, V \otimes \mathbb{C}_\chi)^H$ in the inner product (2.4.5.2). The *normalized induction* of π from H to G is the representation $\mathrm{Ind}_H^G(\pi)$ of G in $\mathrm{Ind}_H^G(V)$ given by

$$(\mathrm{Ind}_H^G(\pi)(g)F)(g') = F(g^{-1}g'),$$

for $g, g' \in G$ and $F \in \mathrm{Ind}_H^G(V)$.

Lemma 2.4.9. *The representation $\mathrm{Ind}_H^G(\pi)$ is continuous and unitary.*

Proof. For continuity, see Lemma 2.22 in [17].

For unitarity, let $F_1, F_2 \in C(G, V \otimes \mathbb{C}_\chi)^H$ and $g \in G$. Then the left-invariance of dg' implies that

$$(\mathrm{Ind}_H^G(\pi)(g)F_1, \mathrm{Ind}_H^G(\pi)(g)F_2) = \int_G \psi(gg')(F_1(g'), F_2(g'))\, dg'.$$

By the last claim in Lemma 2.4.6, this equals (F_1, F_2). □

Example 2.4.10. For a trivial example, suppose that $H = G$. Then evaluation at the identity element defines a unitary equivalence $\mathrm{Ind}_G^G(V) \cong V$. So $\mathrm{Ind}_G^G(\pi) = \pi$.

Example 2.4.11. At the opposite extreme from Example 2.4.10, suppose that $H = \{e\}$ is the trivial subgroup of G. Its only irreducible representation is the trivial representation 1, and $\mathrm{Ind}_{\{e\}}^G(1)$ is the left regular representation of G in $L^2(G)$.

In the situation where we will apply normalized induction of representations, there is an alternative expression for the inner product (2.4.5.2).

Lemma 2.4.12. *In the setting of Lemma 2.4.4, we have for all $F_1, F_2 \in C(G, V \otimes \mathbb{C}_\chi)^H$,*

$$(F_1, F_2) = \int_{H'} (F_1(h'), F_2(h'))_V\, dh'. \qquad (2.4.12.1)$$

This follows from Lemma 2.4.4.

We will also apply normalized induction to non unitary representations π, always in the setting of Lemma 2.4.12. Then the inner product on $C(G, V \otimes \mathbb{C}_\chi)^H$ is defined by (2.4.12.1), because (2.4.5.2) may depend on ψ.

2.4.13 The Discrete Series of Disconnected Groups

Theorem 2.3.37 is the classification of discrete series representations of *connected*, semisimple, linear Lie groups. In the classification of tempered representations, we will also use the discrete series of certain disconnected groups. These can be classified as follows.

Let $G < \mathrm{GL}(n,\mathbb{C}) < \mathrm{GL}(2n,\mathbb{R})$ be a linear Lie group. Suppose that

- G is closed under the conjugate transpose operation;
- G has finitely many connected components;
- the connected component G_0 of the identity element has a compact center;
- if $G^{\mathbb{C}} \subset \mathrm{GL}(2n,\mathbb{C})$ is the group generated by $\exp(\mathfrak{g} \otimes \mathbb{C})$, and $Z_{\mathrm{GL}(2n,\mathbb{C})}(G^{\mathbb{C}})$ is its centralizer in $\mathrm{GL}(2n,\mathbb{C})$, then

$$G \subset G^{\mathbb{C}} Z_{\mathrm{GL}(2n,\mathbb{C})}(G^{\mathbb{C}}).$$

Suppose that G has a compact Cartan subgroup T. Let $\pi_\lambda^{G_0}$ be a discrete series representation of G_0, with corresponding parameter $\lambda \in it^*$ as in Theorem 2.3.37. Let $Z(G)$ be the center of G, and let $\xi \in \widehat{Z(G)}$ be such that

$$\xi|_{T \cap Z(G)} = e^{\lambda - \rho_\lambda^G}|_{T \cap Z(G)}.$$

Then setting

$$(\pi_\lambda^{G_0} \boxtimes \xi)(g_0 z) := \xi(z)\pi_\lambda^{G_0}(g_0)$$

for $g_0 \in G_0$ and $z \in Z(G)$, gives a well-defined representation $\pi_\lambda^{G_0} \boxtimes \xi$ of $G_0 Z(G)$. We write

$$\pi_{\lambda,\xi}^G := \mathrm{Ind}_{G_0 Z(G)}^G(\pi_\lambda^{G_0} \boxtimes \xi).$$

Theorem 2.4.14. *The representation $\pi_{\lambda,\xi}^G$ belongs to the discrete series of G for all $\pi_\lambda^{G_0}$ and ξ as above. All discrete series representations of G occur in this way. Two such representations $\pi_{\lambda,\xi}^G$ and $\pi_{\lambda',\xi'}^G$ are equivalent if and only if $\xi = \xi'$, and there is a $w \in W(G,T)$ such that $\lambda' = w \cdot \lambda$.*

See Section 4 of [19] or [14].

2.4.15 Cuspidal Parabolic Subgroups

We return to the setting where G is a connected, semisimple, linear Lie group. We will apply normalized induction to representations of *cuspidal parabolic subgroups* of G. Intuitively, such a subgroup $P < G$ has a decomposition $P = MAN$, where

- M is "of the same type" as G, though it may be disconnected, and always has discrete series representations;
- A is isomorphic to a vector space;
- N is nilpotent.

Let $H < G$ be a θ-stable Cartan subgroup. Define $\mathfrak{a} := \mathfrak{h} \cap \mathfrak{s}$, with $\mathfrak{s}$ as in (2.2.24.1), and

$$A := \exp(\mathfrak{a}).$$

Then the exponential map defines an isomorphism $A \cong \mathfrak{a}$ as connected, abelian Lie groups.

Let

- $\mathfrak{m}$ be the orthogonal complement to $\mathfrak{a}$ in $Z_{\mathfrak{g}}(\mathfrak{a})$ with respect to (2.2.37.2);
- $M_0 \subset G$ be the subgroup generated by $\exp(\mathfrak{m})$;
- $M := Z_K(\mathfrak{a})M_0 \subset G$.

Lemma 2.4.16. *The subset $M \subset G$ is a subgroup, and it satisfies the assumptions on the group G in Subsection 2.4.13. It has a compact Cartan subgroup.*

See Lemma 1.3 in [19] for the first claim, and Proposition 7.87 in [21] for the last claim.

Choose a positive restricted root system $\Sigma^+(\mathfrak{g},\mathfrak{a})$ for $(\mathfrak{g},\mathfrak{a})$. Let $\mathfrak{n} \subset \mathfrak{g}$ be the corresponding nilpotent subalgebra (2.2.38.1). Let $N < G$ be the subgroup generated by $\exp(\mathfrak{n})$.

Lemma 2.4.17. *(a) The elements of M and A commute.*
(b) The group MA normalizes N.
(c) The groups M, A, and N as above have trivial intersections, and the multiplication map is a diffeomorphism

$$M \times A \times N \to MAN.$$

Part (a) is immediate from the definitions. Parts (b) and (c) are parts (a) and (d) of Proposition 7.83 in [21], respectively.

Because of parts (a) and (b) of Lemma 2.4.17, the subset $MAN < G$ is a subgroup. Furthermore, $MA < MAN$ is a normal subgroup, with quotient N.

Definition 2.4.18. A *cuspidal parabolic subgroup* of G is a subgroup conjugate to a subgroup of the form $P = MAN$, where M, A, and N are as above. The subgroup $MA < P$ is the *Levi factor* of the cuspidal parabolic subgroup P.

Remark 2.4.19. *There is a more general notion of a parabolic subgroup, which we will not use. (See Section VII.7 in [21].) Then the term cuspidal means that the factor M of such a parabolic subgroup has a compact Cartan subgroup, and hence discrete series representations.*

Proposition 2.4.20. *(a) There are finitely many conjugacy classes of cuspidal parabolic subgroups.*
(b) For any cuspidal parabolic subgroup $P < G$, we have $G = KP$.

Part (a) follows from Proposition 7.76 in [21]. Part (b) is Proposition 7.83(f) in [21].

2.4.21 Examples of Cuspidal Parabolic Subgroups

Example 2.4.22. Suppose that $H = T < K$ is a compact Cartan subgroup of G. Then

- $\mathfrak{a} = \{0\}$ and $A = \{e\}$;
- $\mathfrak{m} = \mathfrak{g}$ and $M_0 = G_0$, the connected component of G containing e;
- $M = G$;
- $N = \{e\}$, because $\Sigma(\mathfrak{g}, \{0\}) = \emptyset$.

So the cuspidal parabolic subgroup corresponding to H is $P = G$. The factor $M = G$ has a compact Cartan subgroup by assumption.

So by Theorem 2.3.35, the group G is a cuspidal parabolic subgroup of itself if and only if it has discrete series representations.

Example 2.4.23. Suppose that $G = \mathrm{SL}(2,\mathbb{R})$. For the compact Cartan subgroup $\mathrm{SO}(2) < G$, we obtain the cuspidal parabolic subgroup $P = G$ as in Example 2.4.22.

Let $H < G$ be the diagonal, noncompact Cartan subgroup from Example 2.2.31. Then $\mathfrak{h} \subset \mathfrak{s}$, so $\mathfrak{a} = \mathfrak{h} = \mathbb{R}Y_2$, with Y_2 as in (2.2.31.1). And

$$A = \exp(\mathfrak{a}) = \left\{ \begin{pmatrix} r & 0 \\ 0 & r^{-1} \end{pmatrix}; r > 0 \right\}.$$

Now $Z_{\mathfrak{g}}(\mathfrak{a}) = \mathfrak{a}$, so $\mathfrak{m} = \{0\}$ and $M_0 = \{I\}$. And $Z_K(\mathfrak{a}) = \{\pm I\}$, so

$$M = \{\pm I\}.$$

This is a disconnected group, and both its irreducible representations belong to the discrete series.

Let $\alpha \in \mathfrak{a}^*$ be defined by $\langle \alpha, Y_2 \rangle = 2$. Then $\Sigma(\mathfrak{g}, \mathfrak{a}) = \{\pm\alpha\}$, and as in Example 2.2.39,

$$\mathfrak{g}_\alpha = \mathbb{R}\begin{pmatrix} 0 & 1 \\ 0 & 0 \end{pmatrix},$$

and

$$\mathfrak{g}_{-\alpha} = \mathbb{R}\begin{pmatrix} 0 & 0 \\ 1 & 0 \end{pmatrix}.$$

Choose $\Sigma^+(\mathfrak{g}, \mathfrak{a}) = \{\alpha\}$. Then $\mathfrak{n} = \mathfrak{g}_\alpha$, and

$$N = \exp(\mathfrak{n}) = \left\{ \begin{pmatrix} 1 & x \\ 0 & 1 \end{pmatrix}; x \in \mathbb{R} \right\}.$$

We obtain the cuspidal parabolic subgroup of upper-triangular matrices

$$\begin{aligned} P = MAN &= \left\{ \pm I \begin{pmatrix} r & 0 \\ 0 & r^{-1} \end{pmatrix} \begin{pmatrix} 1 & x \\ 0 & 1 \end{pmatrix}; r > 0, x \in \mathbb{R} \right\} \\ &= \left\{ \begin{pmatrix} y & x \\ 0 & y^{-1} \end{pmatrix}; y \neq 0, x \in \mathbb{R} \right\}. \end{aligned}$$

Example 2.4.24. Suppose that $G = \mathrm{SL}(3, \mathbb{R})$. Then G has two conjugacy classes of Cartan subgroups. One of these is represented by the Cartan subgroup of diagonal matrices. For this Cartan, we obtain, analogously to Example 2.4.23,

$$A = \exp(\mathfrak{h}) = \left\{ \begin{pmatrix} r_1 & 0 & 0 \\ 0 & r_2 & 0 \\ 0 & 0 & (r_1 r_2)^{-1} \end{pmatrix}; r_1, r_2 > 0 \right\}.$$

And

$$M = Z_K(\mathfrak{a}) = \left\{ \begin{pmatrix} \varepsilon_1 & 0 & 0 \\ 0 & \varepsilon_2 & 0 \\ 0 & 0 & \varepsilon_1 \varepsilon_2 \end{pmatrix}; \varepsilon_1, \varepsilon_2 \in \{\pm 1\} \right\}.$$

As in Example 2.2.39, we can choose positive restricted roots such that N consists of the upper-triangular matrices with ones on the diagonal. Then $P = MAN$ consists of all upper-triangular matrices in $\mathrm{SL}(3, \mathbb{R})$ (this computation for the diagonal Cartan extends to $\mathrm{SL}(n, \mathbb{R})$ in general).

Example 2.4.25. The other conjugacy class of Cartan subgroups of $G = \mathrm{SL}(3,\mathbb{R})$ is represented by

$$H := \left\{ \begin{pmatrix} x\cos t & -x\sin t & 0 \\ -x\sin t & x\cos t & 0 \\ 0 & 0 & x^{-2} \end{pmatrix} ; t \in \mathbb{R}, x \in \mathbb{R} \setminus \{0\} \right\}.$$

For this Cartan subgroup,

$$\mathfrak{h} = \mathbb{R} \begin{pmatrix} 0 & -1 & 0 \\ 1 & 0 & 0 \\ 0 & 0 & 0 \end{pmatrix} \oplus \mathbb{R} \begin{pmatrix} 1 & 0 & 0 \\ 0 & 1 & 0 \\ 0 & 0 & -2 \end{pmatrix}.$$

The first summand lies in $\mathfrak{k}$, the second lies in $\mathfrak{s}$. So $\mathfrak{a}$ is spanned by

$$\begin{pmatrix} 1 & 0 & 0 \\ 0 & 1 & 0 \\ 0 & 0 & -2 \end{pmatrix}, \tag{2.4.25.1}$$

and

$$A = \exp(\mathfrak{a}) = \left\{ \begin{pmatrix} r & 0 & 0 \\ 0 & r & 0 \\ 0 & 0 & r^{-2} \end{pmatrix} ; r > 0 \right\}.$$

And

$$Z_{\mathfrak{g}}(\mathfrak{a}) = \left\{ \begin{pmatrix} X & 0 \\ 0 & 0 \end{pmatrix} ; X \in \mathfrak{sl}(2,\mathbb{R}) \right\} \oplus \mathfrak{a}.$$

It folllows that

$$\mathfrak{m} = \left\{ \begin{pmatrix} X & 0 \\ 0 & 0 \end{pmatrix} ; X \in \mathfrak{sl}(2,\mathbb{R}) \right\},$$

and

$$M_0 = \mathrm{SL}(2,\mathbb{R}) \times \{1\},$$

the group generated by $\exp(\mathfrak{m})$.

Furthermore, $K = \mathrm{SO}(3)$, and

$$Z_K(\mathfrak{a}) = \left\{ \begin{pmatrix} g & 0 \\ 0 & \det(g)^{-1} \end{pmatrix} ; g \in \mathrm{O}(2) \right\}.$$

So

$$M = Z_K(\mathfrak{a}) M_0 = \left\{ \begin{pmatrix} g & 0 \\ 0 & \det(g)^{-1} \end{pmatrix} ; g \in \mathrm{SL}(2,\mathbb{R})_{\pm} \right\}.$$

Here

$$\mathrm{SL}(2,\mathbb{R})_{\pm} = \{g \in \mathrm{GL}(2,\mathbb{R}); \det(g) \in \{\pm 1\}\}.$$

Again, M is disconnected.

Let $\alpha \in \mathfrak{a}^*$ be such that its pairing with (2.4.25.1) is 2. Then $\Sigma(\mathfrak{a},\mathfrak{g}) = \{\pm\alpha\}$. And

$$\mathfrak{g}_{\alpha} = \left\{ \begin{pmatrix} 0 & 0 & x \\ 0 & 0 & y \\ 0 & 0 & 0 \end{pmatrix}; x, y \in \mathbb{R} \right\}$$

and

$$\mathfrak{g}_{-\alpha} = \left\{ \begin{pmatrix} 0 & 0 & 0 \\ 0 & 0 & 0 \\ x & y & 0 \end{pmatrix}; x, y \in \mathbb{R} \right\}.$$

Choose $\Sigma^+(\mathfrak{g},\mathfrak{a}) = \{\alpha\}$. Then $\mathfrak{n} = \mathfrak{g}_{\alpha}$, and

$$N = \exp(\mathfrak{n}) = \left\{ \begin{pmatrix} 1 & 0 & x \\ 0 & 1 & y \\ 0 & 0 & 1 \end{pmatrix}; x, y \in \mathbb{R} \right\}.$$

We find the cuspidal parabolic subgroup

$$\begin{aligned} P = MAN &= \left\{ \begin{pmatrix} \multicolumn{2}{c}{\multirow{2}{*}{g}} & 0 \\ & & 0 \\ 0 & 0 & \det(g)^{-1} \end{pmatrix} \begin{pmatrix} r & 0 & 0 \\ 0 & r & 0 \\ 0 & 0 & r^{-2} \end{pmatrix} \begin{pmatrix} 1 & 0 & x \\ 0 & 1 & y \\ 0 & 0 & 1 \end{pmatrix}; \right. \\ &\qquad \left. g \in \mathrm{SL}(2,\mathbb{R})_{\pm}, r > 0, x, y \in \mathbb{R} \right\} \\ &= \left\{ \left(\begin{array}{cc|c} * & * & * \\ * & * & * \\ \hline 0 & 0 & * \end{array} \right) \right\} \subset \mathrm{SL}(3,\mathbb{R}). \end{aligned}$$

Example 2.4.26. Let $G = \mathrm{SL}(n,\mathbb{C})$. Like every complex group, it has only one conjugacy class of Cartan subgroups. We take the diagonal Cartan subgroup

$$H = \left\{ \begin{pmatrix} z_1 & 0 & \cdots & 0 \\ 0 & z_2 & \cdots & 0 \\ \vdots & & \ddots & \vdots \\ 0 & \cdots & 0 & z_n \end{pmatrix}; z_j \in \mathrm{GL}(1,\mathbb{C}), z_1 \cdots z_n = 1 \right\}.$$

Then

$$A = \left\{ \begin{pmatrix} r_1 & 0 & \cdots & 0 \\ 0 & r_2 & \cdots & 0 \\ \vdots & & \ddots & \vdots \\ 0 & \cdots & 0 & r_n \end{pmatrix} ; r_j > 0, r_1 \cdots r_n = 1 \right\} \cong \mathbb{R}^{n-1};$$

$$M = \left\{ \begin{pmatrix} e^{i\alpha_1} & 0 & \cdots & 0 \\ 0 & e^{i\alpha_2} & \cdots & 0 \\ \vdots & & \ddots & \vdots \\ 0 & \cdots & 0 & e^{i\alpha_n} \end{pmatrix} ; \alpha_j \in \mathbb{R}, \alpha_1 + \cdots + \alpha_n = 0 \right\} \cong \mathrm{U}(1)^{n-1}. \tag{2.4.26.1}$$

For the choice of positive restricted roots as in Example 2.2.39, we have

$$N = \left\{ \begin{pmatrix} 1 & x_{12} & \cdots & x_{1n} \\ 0 & 1 & \cdots & x_{2n} \\ \vdots & \ddots & \ddots & \vdots \\ 0 & \cdots & 0 & 1 \end{pmatrix} ; x_{jk} \in \mathbb{C} \right\}.$$

So $P = MAN$ is the group of upper triangular matrices in $\mathrm{SL}(n,\mathbb{C})$.

In general, for a complex semisimple Lie group, there is only one conjugacy class of cuspidal parabolic subgroups, and the factor M is a torus.

2.4.27 Normalized Induction From Cuspidal Parabolic Subgroups

Let G be a connected, semisimple, linear Lie group. Fix a cuspidal parabolic subgroup $P = MAN < G$. If σ is a representation of M in a Hilbert space V, and ν is an irreducible (hence one-dimensional) representation of A, then the map

$$\sigma \otimes \nu \otimes 1_N : P \to \mathrm{GL}(V) \tag{2.4.27.1}$$

given by

$$(\sigma \otimes \nu \otimes 1_N)(man) = \sigma(m)\nu(a),$$

for $m \in M$, $a \in A$, and $n \in N$, is well-defined by part (c) of Lemma 2.4.17. It is a group homomorphism, and hence a representation of P, because of parts (a) and (b) of Lemma 2.4.17. It is a unitary representation if σ and ν are unitary.

Applying normalized induction to such representations (2.4.27.1), as discussed in Subsection 2.4.5, we obtain representations $\mathrm{Ind}_P^G(\sigma \otimes \nu \otimes 1_N)$ of G. These play an important role in the classification of tempered representations, and in other areas of the representation theory of semisimple Lie groups. The following fact is crucial for our discussion.

Theorem 2.4.28. *An irreducible, unitary representation π of G is tempered if and only if there is a cuspidal parabolic subgroup $P = MAN < G$, a discrete series representation $\sigma \in \hat{M}$, and $\nu \in \hat{A}$, such that π is unitarily equivalent to a subrepresentation of* $\mathrm{Ind}_P^G(\sigma \otimes \nu \otimes 1_N)$.

See Theorem 8.53 and Corollary 9.2 in [20].

In the rest of this subsection, we give three descriptions of representations induced from cuspidal parabolic subgroups that are useful in different situations.

For the rest of this subsection, we fix $\sigma \in \hat{M}$ and an irreducible representation ν of A. Let $\Sigma^+(\mathfrak{g}, \mathfrak{a})$ be the positive restricted root system chosen in the construction of N. We will use the element

$$\rho := \frac{1}{2} \sum_{\alpha \in \Sigma^+(\mathfrak{g},\mathfrak{a})} \dim(\mathfrak{g}_\alpha)\alpha \quad \in \mathfrak{a}^*.$$

Because the kernel of the exponential map of A is trivial, this element defines a homomorphism

$$e^\rho : A \to (0, \infty)$$

analogously to (2.3.17.1).

We first describe the *induced picture* of $\mathrm{Ind}_P^G(\sigma \otimes \nu \otimes 1_N)$. We write $\mathbb{C}_\nu$ for the complex numbers equipped with the representation ν of A.

Lemma 2.4.29. *The space $\mathrm{Ind}_P^G(V_\sigma \otimes \mathbb{C}_\nu)$ is the completion of the space of continuous functions $F : G \to V_\sigma$ such that for all $g \in G$, $m \in M$, $a \in A$, and $n \in N$,*

$$F(gman) = e^{-\rho}(a)\nu(a)^{-1}\sigma(m)^{-1}F(g)$$

in the inner product

$$(F_1, F_2) = \int_K (F_1(k), F_2(k))_{V_\sigma}\, dk. \tag{2.4.29.1}$$

Proof. In this setting, $\Delta_G = 1$, and (2.4.2.1) implies that for $m \in M$, $a \in A$, and $n \in N$,

$$\Delta_P(man) = |\det(\mathrm{Ad}_P(man))|^{-1}, \tag{2.4.29.2}$$

where Ad_P is the adjoint action by P on its Lie algebra $\mathfrak{p} = \mathfrak{m} \oplus \mathfrak{a} \oplus \mathfrak{n}$. Part (a) of Lemma 2.4.17 implies that $\mathrm{Ad}(a)$ acts as the identity on $\mathfrak{m} \oplus \mathfrak{a}$. The operator $\mathrm{Ad}(m)$ on $\mathfrak{p}$ has determinant 1 by Corollary 8.30(b) in [21]. And $\mathrm{Ad}(n)$ acts on $\mathfrak{p}$ as a unipotent operator, so the absolute value of its determinant is 1. We find that the right-hand side of (2.4.29.2) equals

$$|\det(\mathrm{Ad}_{\mathfrak{n}}(a))|^{-1} = e^{-2\rho},$$

where $\mathrm{Ad}_{\mathfrak{n}}(a)$ is the adjoint action by a on $\mathfrak{n}$. So

$$\chi(man) = e^{\rho}(a).$$

Together with Lemma 2.4.12 and part (b) of Proposition 2.4.20, this implies the claim. □

Because of part (b) of Proposition 2.4.20, the restriction map from G to K is injective on $C(G, V_\sigma \otimes \mathbb{C}_\chi)$. The image of this restriction map is the space $C(K, V_\sigma)^{K\cap M}$ of continuous functions $F\colon K \to V_\sigma$ such that for all $k \in K$ and $m \in K \cap M$,

$$F(km) = \sigma(m)^{-1}F(k).$$

The restriction map to K is clearly an isometry with respect to the inner product (2.4.29.1) on both spaces $C(G, V_\sigma \otimes \mathbb{C}_\chi)$ and $C(K, V_\sigma)^{K\cap M}$. So $\mathrm{Ind}_P^G(V_\sigma \otimes \mathbb{C}_\nu)$ is unitarily isomorphic to the completion of $C(K, V_\sigma)^{K\cap M}$ in this inner product.

Under this isomorphism, the representation $\mathrm{Ind}_P^G(\sigma \otimes \nu \otimes 1_N)$ of G in $\mathrm{Ind}_P^G(V_\sigma \otimes \mathbb{C}_\nu)$ can be described as follows. For $g \in G$ and $k' \in K$, write

$$g^{-1}k' = kman, \tag{2.4.29.3}$$

for $k \in K$, $m \in M$, $a \in A$, and $n \in N$; this is possible (in a non unique way) by part (b) of Proposition 2.4.20. Consider the representation of G in $C(K, V_\sigma)^{K\cap M}$ given by

$$(g \cdot F)(k') = e^{-\rho}(a)\nu(a)^{-1}\sigma(m)^{-1}F(k),$$

for $g \in G$, $k' \in K$, $F \in C(K, V_\sigma)^{K\cap M}$, and k, m, a, and n as in (2.4.29.3). It extends continuously to the completion of $C(K, V_\sigma)^{K\cap M}$ in the inner product (2.4.29.1), and this representation is unitarily equivalent to $\mathrm{Ind}_P^G(\sigma \otimes \nu \otimes 1_N)$ via the restriction map to K.

This description of $\mathrm{Ind}_P^G(\sigma \otimes \nu \otimes 1_N)$ is the *compact picture* of this representation. Its advantage is that the representation space does not depend on ν.

The third description of $\mathrm{Ind}_P^G(\sigma \otimes \nu \otimes 1_N)$ that we discuss is the *noncompact picture*. Let

$$\bar{N} := \{(n^*)^{-1}; n \in N\}. \tag{2.4.29.4}$$

Lemma 2.4.30. *The subset $\bar{N}MAN \subset G$ is open and dense.*

The proof is analogous to the proof of Theorem 7.66 in [21].

Identifying ν with an element of $\mathfrak{a}^* \otimes \mathbb{C}$ as in (2.3.17.1), we can take its real part. Let φ be the function on $\bar{N}$ given by

$$\varphi(km\exp(X)n) = e^{2\,\mathrm{Re}(\langle \nu, X\rangle)}$$

for $k \in K$, $m \in M$, $X \in \mathfrak{a}$, $n \in N$ such that $km\exp(X)n \in \bar{N}$. Note that the function φ is constant 1 if ν is unitary.

Because of Lemma 2.4.30, the restriction map from G to $\bar{N}$ is injective on $C(G, V_\sigma \otimes \mathbb{C}_\chi)$. This induces a unitary isomorphism from $\mathrm{Ind}_P^G(V_\sigma \otimes \mathbb{C}_\nu)$ onto the space $L^2(\bar{N}, \varphi d\bar{n})$, for a Haar measure $d\bar{n}$ on $\bar{N}$. The representation of G in the latter space corresponding to $\mathrm{Ind}_P^G(\sigma \otimes \nu \otimes 1_N)$ is the following. For $g \in G$ and $\bar{n}' \in \bar{N}$ such that $g^{-1}\bar{n}' \in \bar{N}MAN$, write

$$g^{-1}\bar{n} = \bar{n}man, \tag{2.4.30.1}$$

$\bar{n} \in \bar{N}$, $m \subset M$, $a \in A$, and $n \in N$. Then

$$(g \cdot F)(\bar{n}') = \nu(a)^{-1} e^{-\rho}(a) \sigma(m)^{-1} F(\bar{n}),$$

for $g \in G$, $\bar{n}' \in \bar{N}$, $F \in L^2(\bar{N}, d\bar{n})$, and $\bar{n}$, m, a, and n as in (2.4.30.1).

An advantage of the noncompact picture of $\mathrm{Ind}_P^G(\sigma \otimes \nu \otimes 1_N)$ is that the definition of representation space does not involve transformation behavior of functions under actions by subgroups. Furthermore, the group $\bar{N}$ is contractible.

We end this subsection with the definition of an important type of parabolically induced representation.

Definition 2.4.31. If $P < G$ is a minimal cuspidal parabolic subgroup, then the representations $\mathrm{Ind}_P^G(\sigma \otimes \nu \otimes 1_N)$, where $\sigma \in \hat{M}$ and $\nu \in \hat{A}$, are the *principal series* of G.

A cuspidal parabolic subgroup is minimal if it is not contained in a strictly larger parabolic subgroup. Then its factor M is compact.

2.4.32 Examples of Parabolically Induced Representations

Example 2.4.33. Let $P = MAN$ be the parabolic subgroup of upper-triangular matrices in $G = \mathrm{SL}(2, \mathbb{R})$ from Example 2.4.23. Then

$$\bar{N} = \left\{ \begin{pmatrix} 1 & 0 \\ x & 1 \end{pmatrix} ; x \in \mathbb{R} \right\}.$$

Let 1_M be the trivial representation of $M = \{\pm I\}$, and σ the nontrivial irreducible representation of M.

The representations $P_\nu^+ := \mathrm{Ind}_P^G(1_M \otimes \nu \otimes 1_N)$, for $\nu \in \hat{A} \cong \mathbb{R}$, form the *spherical principal series* of $\mathrm{SL}(2, \mathbb{R})$. Via the noncompact picture, the representation space $\mathrm{Ind}_P^G(V_{1_M} \otimes \mathbb{C}_\nu)$ is unitarily isomorphic to $L^2(\mathbb{R})$. In this realization of $\mathrm{Ind}_P^G(1_M \otimes \nu \otimes 1_N)$, an element $g = \begin{pmatrix} a & b \\ c & d \end{pmatrix} \in \mathrm{SL}(2, \mathbb{R})$ acts on $F \in L^2(\mathbb{R})$ as

$$(P_\nu^+(g)F)(x) = |-bx + d|^{-1-i\nu} F\left(\frac{ax-c}{-bx+d}\right),$$

for $x \in \mathbb{R}$. Here we identified $\hat{A} \cong i\mathfrak{a}^* \cong \mathbb{R}$ via the basis element (2.2.31.1) of $\mathfrak{a}$.

The representations $P_\nu^- := \mathrm{Ind}_P^G(\sigma \otimes \nu \otimes 1_N)$, for $\nu \in \hat{A} \cong \mathbb{R}$, are the *nonspherical principal series* of SL(2,ℝ). Again, the representation space is $L^2(\mathbb{R})$ in the noncompact picture. And with notation as for the spherical principal series, the representation P_ν^- is given by

$$(P_\nu^-(g)F)(x) = \mathrm{sign}(-bx+d)|-bx + d|^{-1-i\nu} F\left(\frac{ax-c}{-bx+d}\right).$$

Example 2.4.34. Let $P = MAN$ be the cuspidal parabolic subgroup of $G = \mathrm{SL}(2,\mathbb{C})$ as in Example 2.4.26. Then

$$\bar{N} = \left\{\begin{pmatrix}1 & 0\\ z & 1\end{pmatrix}; z \in \mathbb{C}\right\}.$$

Now $M \cong \mathrm{U}(1)$ and $A \cong \mathbb{R}$, so $\hat{M} \cong \mathbb{Z}$ and $\hat{A} \cong \mathbb{R}$. For $\sigma \in \hat{M}$ and $\nu \in \hat{A}$, we have the principal series representation $\mathrm{Ind}_P^G(\sigma \otimes \nu \otimes 1_N)$ of SL(2,ℂ). In the noncompact picture, the representation space is $L^2(\mathbb{C})$, and the representation is given by

$$(g \cdot F)(z) = |-bz + d|^{-2-i\nu} \left(\frac{-bz+d}{|-bz+d|}\right)^{-\sigma} F\left(\frac{az-c}{-bz+d}\right),$$

for $g = \begin{pmatrix}a & b\\ c & d\end{pmatrix} \in \mathrm{SL}(2,\mathbb{C})$, $F \in L^2(\mathbb{C})$ and $z \in \mathbb{C}$.

2.5 Intertwining Operators

We have so far seen the classification of discrete series representations in Theorems 2.3.37 and 2.4.14, and seen how unitary representations of the Levi factor MA of a cuspidal parabolic subgroup of a connected, semisimple, linear Lie group G induce unitary representations of G. Our goal is to classify almost all tempered representations of such a group G. By Theorem 2.4.28, it is enough to look inside representations induced from cuspidal parabolic subgroups. Some natural questions are:

1. Which parabolically induced representations are equivalent?
2. Which parabolically induced representations are irreducible?
3. To what extent does a representation induced from a cuspidal parabolic subgroup MAN depend on the choice of N?

Intertwining operators (see Definition 2.3.7) are tools for answering these questions. They can also be used to answer a refinement of the second question: How do parabolically induced representations decompose into irreducible representations? This refinement is discussed in Subsection 2.5.27.

2.5.1 Schur's Lemma

The reason why intertwining operators are helpful in the study of (ir)reducibility of representations is *Schur's lemma*. This is a basic fact that is used in many places.

Let G be a locally compact topological group. For two unitary representations π and π' of G, we write

$$\mathscr{B}(V_\pi, V_{\pi'})^G := \{T \in \mathscr{B}(V_\pi, V_{\pi'}); \pi'(g) \circ T = T \circ \pi(g) \text{ for all } g \in G\}$$

for the space of intertwining bounded operators from V_π to $V_{\pi'}$. If $\pi = \pi'$, then we write $\mathscr{B}(V_\pi)^G := \mathscr{B}(V_\pi, V_\pi)^G$. This is an algebra: the *algebra of intertwiners* of π.

Lemma 2.5.2 (Schur). *(a) If π is a unitary representation of G, then π is irreducible if and only if $\mathscr{B}(V_\pi)^G = \mathbb{C}\mathrm{Id}_{V_\pi}$, where Id_{V_π} is the identity operator on V_π.*
(b) If $\pi, \pi' \in \hat{G}$, then

$$\dim \mathscr{B}(V_\pi, V_{\pi'})^G = \begin{cases} 0 & \text{if } \pi \not\cong \pi'; \\ 1 & \text{if } \pi \cong \pi'. \end{cases}$$

Proof. See Proposition 1.5 in [20].

In the finite-dimensional case, a nonzero $T \in \mathscr{B}(V_\pi, V_{\pi'})^G$ is injective because its kernel is a G-invariant proper subspace of V_π, and surjective because its image is a G-invariant nonzero subspace of $V_{\pi'}$. So T is a G-equivariant isomorphism, and we may assume that $\pi = \pi'$. In that case, for any eigenvalue λ of T, the map $T - \lambda \mathrm{Id}_{V_\pi}$ is not an isomorphism, and hence the zero map. □

Corollary 2.5.3. *Let $\pi_1, \ldots, \pi_n \in \hat{G}$ be mutually inequivalent. Let $m_1, \ldots, m_n \in \mathbb{N}$, and consider the unitary representation*

$$\pi = \bigoplus_{j=1}^{n} m_j \pi_j$$

of G, where (as in (2.3.14.1)), $m_j \pi_j$ is the direct sum of m_j copies of π_j. Then

$$\mathscr{B}(V_\pi)^G \cong \bigoplus_{j=1}^{n} M_{m_j}(\mathbb{C}).$$

In particular, $\mathscr{B}(V_\pi)^G$ is commutative if and only if $m_j = 1$ for all j.

The moral is that if $\mathscr{B}(V_\pi)^G$ is finite-dimensional, for a unitary representation π, then this algebra determines how many irreducible representations occur in its decomposition, and how often each occurs.

2.5.4 A Weyl Group

For the rest of this section, we suppose that G is a connected, semisimple, linear Lie group. Let $P = MAN < G$ be a cuspidal parabolic subgroup. The adjoint action by M on $\mathfrak{g}$ is trivial on $\mathfrak{a}$ by part (a) of Lemma 3.10. So the adjoint representation defines an embedding of $K \cap M$ as a normal subgroup of

$$N_K(\mathfrak{a}) := \{k \in K; \mathrm{Ad}(k)\mathfrak{a} \subset \mathfrak{a}\}.$$

We will use the Weyl group

$$W := N_K(\mathfrak{a})/(K \cap M).$$

Lemma 2.5.5. *The group W is finite.*

Proof. By Proposition 7.78(c) and Corollary 7.81 in [21], the Lie algebra of W is trivial. □

For all $k \in N_K(\mathfrak{a})$, conjugation by k preserves $Z_{\mathfrak{g}}(\mathfrak{a})$ and $Z_K(\mathfrak{a})$, and hence M. So for a representation σ of M, we have the representation $k \cdot \sigma$ of M, given by

$$(k \cdot \sigma)(m) = \sigma(k^{-1}mk), \tag{2.5.5.1}$$

for $m \in M$. Similarly, for a representation ν of A, we can define the representation $k \cdot \nu$ of A by

$$(k \cdot \nu)(a) = \nu(k^{-1}ak), \tag{2.5.5.2}$$

for $a \in A$.

If $k \in K \cap M < N_K(\mathfrak{a})$, then the conjugation action by k on A is trivial. So $k \cdot \nu = \nu$ for all representations ν of A. And if σ is a unitary representation of M, then

$$(k \cdot \sigma)(m) = \sigma(k)^{-1}\sigma(m)\sigma(k)$$

for all $m \in M$, so the operator $\sigma(k)\colon V_\sigma \to V_\sigma$ is a unitary equivalence between $k \cdot \sigma$ and σ.

We find that (2.5.5.1) and (2.5.5.2) define an action by $N_K(\mathfrak{a})$ on $\hat{M} \times \hat{A}$, such that the normal subgroup $K \cap M$ acts trivially. Hence the action descends to an action by W on $\hat{M} \times \hat{A}$.

The group W also permutes the possible nilpotent components N of cuspidal parabolic subgroups with Levi factor MA. Indeed, if $k \in N_K(\mathfrak{a})$, then a short computation shows that for all $\alpha \in \Sigma(\mathfrak{g}, \mathfrak{a})$,

$$\mathrm{Ad}(k)\mathfrak{g}_\alpha = \mathfrak{g}_{\alpha \circ \mathrm{Ad}(k)^{-1}}. \tag{2.5.5.3}$$

So $\alpha \circ \mathrm{Ad}(k)^{-1} \in \Sigma(\mathfrak{g}, \mathfrak{a})$. And the subgroup group $K \cap M$ acts trivially on $\mathfrak{a}$ and hence on $\Sigma(\mathfrak{g}, \mathfrak{a})$. Hence, we obtain an action by W on $\Sigma(\mathfrak{g}, \mathfrak{a})$.

Because $K \cap M$ normalizes N, we can unambiguously define the nilpotent subgroup $wNw^{-1} < G$, for $w \in W$. If N is associated with the positive restricted root system $\Sigma^+(\mathfrak{g}, \mathfrak{a})$, then (2.5.5.3) implies that wNw^{-1} is the nilpotent subgroup of G defined by the positive restricted root system $w \cdot \Sigma^+(\mathfrak{g}, \mathfrak{a})$.

Example 2.5.6. Suppose that G has discrete series representations, and that $G = P$. Then $\mathfrak{a} = \{0\}$ and $M = G$, so $W = K/K$ is the trivial group.

Example 2.5.7. Suppose that $G = \mathrm{SL}(2, \mathbb{R})$, and let $P = MAN$ be the cuspidal parabolic subgroup of upper-triangular matrices, as in Example 2.4.23. Then

$$N_K(\mathfrak{a}) = \left\{ k \in \mathrm{SO}(2); k \begin{pmatrix} 1 & 0 \\ 0 & -1 \end{pmatrix} k^{-1} \in \mathbb{R} \begin{pmatrix} 1 & 0 \\ 0 & -1 \end{pmatrix} \right\} = \left\{ \pm I, \pm \begin{pmatrix} 0 & -1 \\ 1 & 0 \end{pmatrix} \right\}.$$

And $K \cap M = \{\pm I\}$, so

$$W = N_K(\mathfrak{a})/(K \cap M) = \left\{ e, \begin{pmatrix} 0 & -1 \\ 1 & 0 \end{pmatrix} M \right\}.$$

Now W acts trivially on M. The adjoint action by the nontrivial element of W on $\mathfrak{a} \cong \mathbb{R}$ is reflection in 0.

2.5.8 Knapp–Stein Intertwining Operators

Again, let $P = MAN < G$ be a cuspidal parabolic subgroup. Recall the definition of the group $\bar{N}$ in (2.4.29.4). Let $w \in W$. The Haar measure dn on N corresponds to a Haar measure on wNw^{-1}; that in turn descends to an wNw^{-1}-invariant measure $d\dot{n}$ on $wNw^{-1}/(wNw^{-1} \cap N)$. The inclusion map $wNw^{-1} \cap \bar{N} \hookrightarrow wNw^{-1}$ induces a diffeomorphism

$$wNw^{-1} \cap \bar{N} \cong wNw^{-1}/(wNw^{-1} \cap N).$$

(See Section VII.4 of [20].) Through this identification, we obtain a measure $d\dot{n}$ on $wNw^{-1} \cap \bar{N}$.

Let $\sigma \in \hat{M}$, and let ν be an irreducible representation of A. For $F \in \mathrm{Ind}_P^G(V_\sigma \otimes \mathbb{C}_\nu)$ and $g \in G$, define

$$(A_P(w, \sigma, \nu)F)(g) := \int_{wNw^{-1} \cap \bar{N}} F(gw\dot{n})\, d\dot{n}, \tag{2.5.8.1}$$

whenever this converges.

The Lie algebra of $\bar{N}$ is

$$\bar{\mathfrak{n}} = \{X^*; X \in \mathfrak{n}\}.$$

Theorem 2.5.9 ([18]). *(a) Let $F \in \operatorname{Ind}_P^G(V_\sigma \otimes \mathbb{C}_\nu)$ be smooth. For all $\alpha \in \Sigma(\mathfrak{g}, \mathfrak{a})$ such that $\mathfrak{g}_\alpha \subset \mathfrak{n} \cap \operatorname{Ad}(w)\bar{\mathfrak{n}}$, there is a $c_\alpha > 0$, such that (2.5.8.1) converges for all ν such that $(\operatorname{Re}(\nu), \alpha) > c_\alpha$ for all such α, where we identify ν with the corresponding element of $\mathfrak{a}^* \otimes \mathbb{C}$.*

(b) If F is K-finite (Definition 2.3.26), then the function

$$\nu \mapsto A_P(w, \sigma, \nu)F \tag{2.5.9.1}$$

has a meromorphic continuation to $\nu \in \mathfrak{a}^* \otimes \mathbb{C}$. Here we use the compact picture of $\operatorname{Ind}_P^G(\sigma \otimes \nu \otimes 1_N)$ to realize the right-hand side of (2.5.9.1) in a space independent of ν.

See Theorem 6.6 in [18].

Because of the condition $(\operatorname{Re}(\nu), \alpha) > c_\alpha$ in part (a) of Theorem 2.5.9, we need to consider nonunitary ν even if our end goal is to study intertwining operators for unitary representations. Because the poles in the meromorphic continuation in part (b) of Theorem 2.5.9 may be located at unitary representations of A, Knapp and Stein used normalizing factors to cancel these poles.

Theorem 2.5.10 ([18]). *There are explicit functions $\nu \mapsto \gamma_P(w, \sigma, \nu)$ on $\mathfrak{a}^* \otimes \mathbb{C}$ with the following properties.*

(a) The operator

$$\mathscr{A}_P(w, \sigma, \nu) := \frac{1}{\gamma_P(w, \sigma, \nu)} A_P(w, \sigma, \nu)$$

on K-finite functions $F \in \operatorname{Ind}_P^G(V_\sigma \otimes \mathbb{C}_\nu)$ extends holomorphically to $\nu \in \hat{A}$.

(b) If $\nu \in \hat{A}$, then $\mathscr{A}_P(w, \sigma, \nu)$ extends to a unitary, intertwining operator

$$\mathscr{A}_P(w, \sigma, \nu) \colon \operatorname{Ind}_P^G(V_\sigma \otimes \mathbb{C}_\nu) \to \operatorname{Ind}_P^G(V_{w \cdot \sigma} \otimes \mathbb{C}_{w \cdot \nu}).$$

(c) For all $w, w' \in W$,

$$\mathscr{A}_P(w, w' \cdot \sigma, w' \cdot \nu)\mathscr{A}_P(w', \sigma, \nu) = \mathscr{A}_P(ww', \sigma, \nu).$$

(d) For all $T \in \mathrm{U}(V_\sigma)$,

$$T\mathscr{A}_P(w, \sigma, \nu)T^{-1} = \mathscr{A}_P(w, T\sigma T^{-1}, \nu).$$

See Proposition 8.6 in [18]. Part (b) of Theorem 2.5.10 is a tool for answering the first question at the start of this section.

For another cuspidal parabolic subgroup of the form $P' = MAN' < G$, with N' associated with another positive restricted root system than N', there are analogous intertwining operators

$$\mathscr{A}(P', P, \sigma, \nu)\colon \mathrm{Ind}_P^G(V_\sigma \otimes \mathbb{C}_\nu) \to \mathrm{Ind}_{P'}^G(V_\sigma \otimes \mathbb{C}_\nu). \tag{2.5.10.1}$$

These can be used to show that $\mathrm{Ind}_P^G(\sigma \otimes \nu \otimes 1_N)$ and $\mathrm{Ind}_{P'}^G(\sigma \otimes \nu \otimes 1_{N'})$ are unitarily equivalent, if $\nu \in \hat{A}$. See Proposition 8.5 in [18]. This answers the third question at the start of this section.

Example 2.5.11. Suppose that $G = \mathrm{SL}(2,\mathbb{R})$. Consider the principal series representations $P_\nu^+ = \mathrm{Ind}_P^G(1_M \otimes \nu \otimes 1_N)$ and $P_\nu^- = \mathrm{Ind}_P^G(\sigma \otimes \nu \otimes 1_N)$ from Example 2.4.33. We saw in Example 2.5.7 that W is the two-element group; let $w \in W$ be the nontrivial element. We also saw that w acts trivially on $\hat{M}$, and as reflection in the origin on $\hat{A} \cong \mathbb{R}$. This yields unitary intertwining operators

$$\begin{aligned} \mathscr{A}_P(w, 1_M, \nu)&\colon \mathrm{Ind}_P^G(\mathbb{C}_\nu) \to \mathrm{Ind}_P^G(\mathbb{C}_{-\nu});\\ \mathscr{A}_P(w, \sigma, \nu)&\colon \mathrm{Ind}_P^G(V_\sigma \otimes \mathbb{C}_\nu) \to \mathrm{Ind}_P^G(V_\sigma \otimes \mathbb{C}_{-\nu}). \end{aligned} \tag{2.5.11.1}$$

This shows that $P_\nu^\pm \cong P_{-\nu}^\pm$ for all ν.

The intertwining operators (2.5.11.1) can be described as follows. If $t \in (0, 1/2)$ and $\nu \in \mathbb{R}$, then the function

$$y \mapsto \frac{1}{|y|^{1-t-i\nu}}$$

is locally integrable on $\mathbb{R}$, and square-integrable on $\mathbb{R}\setminus[-1,1]$. So for all smooth functions $F \in L^2(\mathbb{R})$ and all $x \in \mathbb{R}$, the integrals

$$\int_{\mathbb{R}} \frac{F(x-y)}{|y|^{1-t-i\nu}}\, dy \quad \text{and} \quad \int_{\mathbb{R}} \frac{F(x-y)}{|y|^{1-t-i\nu}} \operatorname{sign}(y)\, dy$$

converge. In the noncompact picture, where the representation spaces of $P_\nu^\pm$ are $L^2(\mathbb{R})$, the intertwining operators (2.5.11.1) are extensions of

$$\begin{aligned} (\mathscr{A}_P(w, 1_M, \nu)F)(x) &= \lim_{t \downarrow 0} \int_{\mathbb{R}} \frac{F(x-y)}{|y|^{1-t-i\nu}}\, dy;\\ (\mathscr{A}_P(w, \sigma, \nu)F)(x) &= \lim_{t \downarrow 0} \int_{\mathbb{R}} \frac{F(x-y)}{|y|^{1-t-i\nu}} \operatorname{sign}(y)\, dy. \end{aligned} \tag{2.5.11.2}$$

2.5.12 Completeness and Irreducibility

We still consider a cuspidal parabolic subgroup $P = MAN < G$.

Lemma 2.5.13. *Let $\sigma \in \hat{M}$ and $k \in N_K(\mathfrak{a})$ be such that $k \cdot \sigma \cong \sigma$. Then the representation σ extends to a unitary representation of the subgroup of G generated by M and k.*

See Lemma 7.9 in [18] and the comments below it. In the situation of Lemma 2.5.13, the definition (2.5.5.1) of the representation $k \cdot \sigma$ implies that the operator

$$\sigma(k)\colon V_\sigma \to V_\sigma$$

intertwines $k \cdot \sigma$ and σ. Together with part (b) of Theorem 2.5.10, this has the following consequence.

Corollary 2.5.14. *Let $\sigma \in \hat{M}$, $\nu \in \hat{A}$, and $w \in W$, and suppose that $w \cdot \sigma \cong \sigma$ and $w \cdot \nu = \nu$. Then the composition*

$$\sigma(w)\mathscr{A}_P(w,\sigma,\nu)\colon \operatorname{Ind}_P^G(V_\sigma \otimes \mathbb{C}_\nu) \to \operatorname{Ind}_P^G(V_\sigma \otimes \mathbb{C}_\nu) \qquad (2.5.14.1)$$

is a unitary intertwining operator.

If $w \in K \cap M$, then the operator (2.5.14.1) is the identity operator. So it indeed makes sense to talk about this operator as depending on $w \in W$ with the properties in Corollary 2.5.14.

Remark 2.5.15. *In Corollary 2.5.14 and other places, it is important to distinguish between two representations (ν and $w \cdot \nu$) being exactly equal and two representations (σ and $w \cdot \sigma$) being unitarily equivalent. This is because the unitary isomorphism ($\sigma(w)$) that implements the unitary equivalence plays a role.*

Corollary 2.5.14 allows us to formulate a result of Harish-Chandra in terms of Knapp–Stein intertwiners. Let $\sigma \in \hat{M}$ and $\nu \in \hat{A}$, and consider the subgroup

$$W_{\sigma,\nu} := \{k \in N_K(\mathfrak{a}); k \cdot \sigma \cong \sigma, k \cdot \nu = \nu\}/(K \cap M) < W. \qquad (2.5.15.1)$$

(As noted below (2.5.5.2), elements $K \cap M$ fix ν exactly, and σ up to equivalence.)

Theorem 2.5.16 (Harish-Chandra's completeness theorem). *If σ belongs to the discrete series of M, then the algebra $\mathscr{B}(\operatorname{Ind}_P^G(V_\sigma \otimes \mathbb{C}_\nu))^G$ of intertwiners of $\operatorname{Ind}_P^G(\sigma \otimes \nu \otimes 1_N)$ is spanned by*

$$\{\sigma(w)\mathscr{A}_P(w,\sigma,\nu); w \in W_{\sigma,\nu}\}.$$

See Theorem 14.31 in [20], Theorem 38.1 in [16], Corollary 9.8 in [18], or Theorem 38.1 and Lemma 38.2 in [29].

Together with Schur's lemma (Lemma 2.5.2), Theorem 2.5.16 implies a criterion for irreducibility.

Corollary 2.5.17. *If σ belongs to the discrete series of M and $W_{\sigma,\nu} = \{e\}$, then* $\mathrm{Ind}_P^G(\sigma \otimes \nu \otimes 1_N)$ *is irreducible.*

The action by W on $\mathfrak{a}$ gives an action by W on $\hat{H} \cong i\mathfrak{a}^*$.

Definition 2.5.18. A representation $\nu \in \hat{A}$ is *regular* if it is not fixed by any nontrival element of W.

All nontrivial elements of W fix a strict linear subspace of $i\mathfrak{a}^*$. So the set of regular representations of $\hat{A}$ is open and dense. Corollary 2.5.17 implies the following. (See Theorem 14.15 in [20] for a proof not based on Theorem 2.5.16.)

Corollary 2.5.19. *If $\sigma \in \hat{M}$ belongs to the discrete series, and $\nu \in \hat{A}$ is regular, then* $\mathrm{Ind}_P^G(\sigma \otimes \nu \otimes 1_N)$ *is irreducible.*

This is an answer to the second question at the start of this section, for induced representations parametrized by an open dense set.

In general, by Lemmas 2.5.2 and 2.5.5, Theorem 2.5.16 implies that $\mathrm{Ind}_P^G(\sigma\otimes\nu\otimes 1_N)$ is a direct sum of finitely many irreducible representations for all $\nu \in \hat{A}$ and all σ in the discrete series of M. Corollary 2.5.34 is a refinement of this statement.

Example 2.5.20. Suppose that G has a compact Cartan subgroup, so that it has discrete series representations. Let π be a discrete series representation of G. For the cuspidal parabolic $G < G$ as in Example 2.4.22, we have $M = G$, and A and N are trivial. And by Example 2.4.10, we have

$$\mathrm{Ind}_G^G(\pi \otimes 1_A \otimes 1_N) = \pi.$$

Now $\mathfrak{a} = \{0\}$, so

$$W = K/K = \{e\}.$$

In particular, $W_{\pi,1_A} = \{e\}$. So Corollary 2.5.19 is consistent with the fact that π is irreducible.

Example 2.5.21. Suppose that $G = \mathrm{SL}(2,\mathbb{R})$, and let $P = MAN$ be the cuspidal parabolic subgroup of upper-triangular matrices, as in Example 2.4.23. We saw in Example 2.5.7 that W has two elements, and acts trivially on M, and by reflection in the origin on $\mathfrak{a} = \mathbb{R}$.

Let $\sigma \in \hat{M}$ and $\nu \in \hat{A}$. Then by the form of the action by W on M and $\mathfrak{a}$,

$$W_{\sigma,\nu} = \begin{cases} \{e\} & \text{if } \nu \neq 1_A; \\ W & \text{if } \nu = 1_A. \end{cases}$$

So Corollary 2.5.17 or Corollary 2.5.19 implies that the principal series representation $\mathrm{Ind}_P^G(\sigma \otimes \nu \otimes 1_N)$ is irreducible if $\nu \neq 1_A$.

If $\nu = 1_A$, then $W_{\sigma,\nu} = W$ has two elements. So Theorem 2.5.16 and Corollary 2.5.3 imply that $\mathrm{Ind}_P^G(\sigma \otimes 1_A \otimes 1_N)$ is the direct sum of at most two irreducible representations, and that those two are then inequivalent. To determine if $\mathrm{Ind}_P^G(\sigma \otimes 1_A \otimes 1_N)$ is irreducible or not, we need more theory; see Example 2.5.37.

2.5.22 Classification of Tempered Representations

We mentioned that Theorem 2.5.10 can be used to find equivalences between induced representations, but more is needed to show that we have found all possible equivalences.

Theorem 2.5.23 (Langlands disjointness theorem). *For $j = 1,2$, consider cuspidal parabolics $P_j = M_jA_jN_j$, and $\sigma_j \in \hat{M}_j$ and $\nu_j \in \hat{A}_j$. If $\mathrm{Ind}_{P_1}^G(\sigma_1 \otimes \nu_1 \otimes 1_N)$ and $\mathrm{Ind}_{P_2}^G(\sigma_2 \otimes \nu_2 \otimes 2)$ have a common irreducible subrepresentation, then there is a $k \in K$ such that*

$$\begin{aligned} M_2 &= kM_1k^{-1} \\ A_2 &= kA_1k^{-1} \\ \sigma_2 &\cong k \cdot \sigma_1 \\ \nu_2 &= k \cdot \nu_1. \end{aligned} \tag{2.5.23.1}$$

See Theorem 14.90 in [20] or Lemma 3.14 in [25].

Combining the results we have collected so far, we obtain a classification of almost all tempered representations.

Theorem 2.5.24. *Let $H_1, \ldots, H_n < G$ be θ-stable representatives of the conjugacy classes of Cartan subgroups of G. Let $P_j = M_jA_jN_j$ be the cuspidal parabolic subgroup associated with H_j and a choice of positive restricted roots. The representations $\mathrm{Ind}_{P_j}^G(\sigma_j \otimes \nu_j \otimes 1_{N_j})$, where*

1. *$\sigma_j \in \hat{M}_j$ belongs to the discrete series and*
2. *$\nu_j \in \hat{A}_j$ is regular,*

are irreducible and tempered. Almost every irreducible tempered representation of G is of this form.

Two such representations $\mathrm{Ind}_{P_j}^G(\sigma_j \otimes \nu_j \otimes 1_{N_j})$ and $\mathrm{Ind}_{P_{j'}}^G(\sigma'_{j'} \otimes \nu'_{j'} \otimes 1_{N_{j'}})$ are equivalent if and only if $j = j'$, and there is a $w \in W_j = N_K(\mathfrak{a}_j)/(K \cap M_j)$ such that $\sigma'_j \cong w \cdot \sigma_j$ and $\nu'_j = w \cdot \nu_j$.

Proof. Let $\pi \in \hat{G}$ be tempered. By Theorem 2.4.28, π lies inside a representation of the form $\mathrm{Ind}_P^G(\sigma \otimes \nu \otimes 1_N)$, for a cuspidal parabolic subgroup $P = MAN$, a discrete series representation σ of M, and $\nu \in \hat{A}$. By Corollary 2.5.19, the representation $\mathrm{Ind}_P^G(\sigma \otimes \nu \otimes 1_N)$ is irreducible if ν lies in the open dense set of regular representations in $\hat{A}$. So if ν is regular, then $\pi = \mathrm{Ind}_P^G(\sigma \otimes \nu \otimes 1_N)$.

The cuspidal parabolic subgroup P is associated with a θ-stable Cartan subgroup H. By Theorem 2.2.26, there are $k \in K$ and j such that $H = kH_jk^{-1}$. So $A = kA_jk^{-1}$, and $M = kM_jk^{-1}$. The nilpotent group $\tilde{N}_j := k^{-1}Nk$ is associated with the positive restricted root system

$$\{\alpha \circ \mathrm{Ad}(k); \alpha \in \Sigma^+(\mathfrak{g}, \mathfrak{a})\}$$

in $\Sigma(\mathfrak{g}, \mathfrak{a}_j)$ that we get by applying k^{-1} to the positive restricted root system $\Sigma^+(\mathfrak{g}, \mathfrak{a}) \subset \Sigma(\mathfrak{g}, \mathfrak{a})$ that defines N. Let $\sigma_j := k^{-1} \cdot \sigma$ and $\nu_j := k^{-1} \cdot \nu \in \hat{A}_j$. Then σ_j belongs to the discrete series of M_j, and

$$\mathrm{Ind}_P^G(\sigma \otimes \nu \otimes 1_N) = k \cdot \mathrm{Ind}_{M_jA_j\tilde{N}_j}^G(\sigma_j \otimes \nu_j \otimes 1_{\tilde{N}_j}),$$

The representation on the right is equivalent to $\mathrm{Ind}_{M_jA_j\tilde{N}_j}^G(\sigma_j \otimes \nu_j \otimes 1_{\tilde{N}_j})$, via the inner automorphism of conjugating by k. The latter representation is equivalent to $\mathrm{Ind}_{P_j}^G(\sigma_j \otimes \nu_j \otimes 1_{N_j})$ via a Knapp–Stein intertwiner of the type (2.5.10.1).

Now suppose that $\mathrm{Ind}_{P_j}^G(\sigma_j \otimes \nu_j \otimes 1_{N_j}) \cong \mathrm{Ind}_{P_{j'}}^G(\sigma'_{j'} \otimes \nu'_{j'} \otimes 1_{N_{j'}})$. By Theorem 2.5.23, there is a $k \in K$ such that

$$\begin{aligned} M_{j'} &= kM_jk^{-1} \\ A_{j'} &= kA_jk^{-1} \\ \sigma'_{j'} &\cong k \cdot \sigma_j \\ \nu'_{j'} &= k \cdot \nu_j. \end{aligned} \tag{2.5.24.1}$$

The second equality implies that

$$H_{j'} = Z_K(A_{j'})A_{j'} = (kZ_K(A_jk^{-1})(kA_{j'}k^{-1}) = kH_jk^{-1}.$$

So $j' = j$. Hence $k \in N_K(\mathfrak{a}_j)$, and the last claim follows. □

Remark 2.5.25. *The meaning of "almost every" in Theorem 2.5.24 is that we obtain all tempered representations in an open dense subset of the parameter space associated with each conjugacy class of Cartan subgroups. It turns out that the Plancherel measure on every connected component of such a parameter space is a function times Lebesgue measure on* $\hat{A}_j \cong \mathfrak{a}_j$. *This implies that we indeed find almost every tempered representation in the measure-theoretic sense. This is enough to compute the Plancherel measure, which was*

Harish-Chandra's goal. The more detailed classification of all tempered representations was completed by Knapp and Zuckerman [22, 23]. This involves the R-groups *discussed in Subsection 2.5.27; see Remark 2.5.38.*

Example 2.5.26. Let $G = \mathrm{SL}(2,\mathbb{R})$. The tempered representations associated with the compact Cartan subgroup SO(2) are the discrete series representations, described in Subsection 2.3.38. The tempered representations associated with the noncompact Cartan of diagonal matrices, and regular $\nu \in \hat{A}$, are the principal series representations $P^{\pm}_{\nu}$ in Example 2.4.33, for $\nu \neq 0$. The only equivalences between these representations are the ones in Example 2.5.11: $P^{\pm}_{\nu} \cong P^{\pm}_{-\nu}$ for all ν. For the case $\nu = 0$, see Example 2.5.37.

2.5.27 R-groups and Reducibility

As before, let G be a connected, semisimple, linear Lie group. Let $P = MAN$ be a cuspidal parabolic subgroup. We fix a discrete series representation σ of M, and $\nu \in \hat{A}$. We saw in Corollary 2.5.19 that the representation $\mathrm{Ind}_P^G(\sigma \otimes \nu \otimes 1_N)$ is irreducible if ν is regular. And by Theorem 2.5.24, this case is enough to classify almost all tempered representations of G, and hence to determine its Plancherel measure.

In this subsection, we focus on the case of singular $\nu \in \hat{A}$. This is relevant for the classification of all tempered representations by Knapp and Zuckerman, and for the description of the reduced group C^*-algebra of G in Subsection 2.6.

By Corollary 2.5.3, the algebra $\mathscr{B}(\mathrm{Ind}_P^G(V_\sigma \otimes \mathbb{C}_\nu))^G$ of intertwiners of $\mathrm{Ind}_P^G(\sigma \otimes \nu \otimes 1_N)$ contains a lot of information on the decomposition into irreducibles of these representations. By Theorem 2.5.16, this algebra is spanned by the operators $\sigma(w)\mathscr{A}_P(w,\sigma,\nu)$, where w runs over the group $W_{\sigma,\nu}$ in (2.5.15.1). For a more explicit description of the algebra of intertwiners, it would be useful to have a *basis*, rather than a spanning set. A first observation is that we may omit the operators $\sigma(w)\mathscr{A}_P(w,\sigma,\nu)$ for nontrivial elements of the subgroup

$$W'_{\sigma,\nu} := \{w \in W_{\sigma,\nu}; \sigma(w)\mathscr{A}_P(w,\sigma,\nu) \in \mathbb{C}\mathrm{Id}\}$$

and still obtain a set spanning the algebra of intertwiners. It turns out that this is enough to obtain a basis.

Theorem 2.5.28 ([18, 19]). *There is a subgroup $R_{\sigma,\nu} < W_{\sigma,\nu}$ such that*

1. $W_{\sigma,\nu} = W'_{\sigma,\nu} \rtimes R_{\sigma,\nu}$;
2. *the set of operators*

$$\{\sigma(w)\mathscr{A}_P(w,\sigma,\nu); w \in R_{\sigma,\nu}\}$$

   is a basis of $\mathscr{B}(\mathrm{Ind}_P^G(V_\sigma \otimes \mathbb{C}_\nu))^G$;
3. *there is an* $n \leq \dim A$ *such that* $R_{\sigma,\nu} \cong (\mathbb{Z}/2\mathbb{Z})^n$.

See Theorem 13.4 in [18] for the first two points, and Theorem 15.1 in [18] and Theorem 6.1 in [19] for the third point.

Definition 2.5.29. The group $R_{\sigma,\nu}$ in Theorem 2.5.28 is the *R-group* associated with σ and ν.

Remark 2.5.30. *The group $R_{\sigma,\nu}$ can be expressed in terms of the Plancherel densities for certain subgroups of G, see Theorem 13.4(ii) and equalities (10.3) and (10.4) in [18]. This expression justifies talking about* the *group $R_{\sigma,\nu}$.*

Remark 2.5.31. *If ν is regular, then $W_{\sigma,\nu}$ is the trivial group, and Theorem 2.5.28 is immediate from Corollary 2.5.19.*

Lemma 2.5.32. *For all $k, k' \in N_K(\mathfrak{a})$ such that $k \cdot \sigma \cong \sigma$ and $k' \cdot \sigma \cong \sigma$, there is a $c \in \mathrm{U}(1)$, determined by*

$$\sigma(k)\sigma(k') = c\sigma(kk'),$$

such that

$$\sigma(k)\mathscr{A}_P(k,\sigma,k' \cdot \nu)\sigma(k')\mathscr{A}_P(k',\sigma,\nu) = c\sigma(kk')\mathscr{A}_P(kk',\sigma,\nu).$$

The case where $k \cdot \nu = k' \cdot \nu = \nu$ is Lemma 14.38 in [20]; the proof generalizes directly to our setting. This is a short argument, based on parts (c) and (d) of Theorem 2.5.10.

Corollary 2.5.33 (Multiplicity one theorem). *The algebra $\mathscr{B}(\mathrm{Ind}_P^G(V_\sigma \otimes \mathbb{C}_\nu))^G$ is commutative.*

See Corollary 14.66 in [20] or Theorem 7.1 in [19]. Key ingredients are part (c) of Theorem 2.5.28, commutativity of $R_{\sigma,\nu}$ (by part(b) of Theorem 2.5.28), and Lemma 2.5.32.

Corollaries 2.5.3 and 2.5.33 and part (c) of Theorem 2.5.28 have the following consequence.

Corollary 2.5.34. *The representation $\mathrm{Ind}_P^G(\sigma \otimes \nu \otimes 1_N)$ decomposes as the direct sum of $\#R_{\sigma,\nu}$ inequivalent irreducible representations.*

For an explicit decomposition of $\mathrm{Ind}_P^G(\sigma \otimes \nu \otimes 1_N)$ into irreducible representations, see Theorem 14.79 in [20] or Theorem 8.7 in [22].

Theorem 2.5.35. *If the dimension of A is the lowest among all θ-stable Cartan subgroups, then $R_{\sigma,\nu}$ is the trivial group.*

See Corollary 14.60 in [20].

Theorem 2.5.35 applies in particular to discrete series representations (where irreducibility was already known) and to complex groups.

Example 2.5.36. If G is a complex, semisimple, connected, linear Lie group such as $\mathrm{SL}(n,\mathbb{C})$, then all its Cartan subgroups are conjugate, and the components M of cuspidal parabolic subgroups are tori. So all irreducible unitary representations of M belong to the discrete series. If $A \cong \mathbb{R}^d$, then $M \cong \mathbb{R}^d/\mathbb{Z}^d$. So $\hat{A} \cong \mathbb{R}^d$, and by Lemma 2.3.17, $\hat{M} \cong \mathbb{Z}^d$. By Theorem 2.5.35, all representations $\mathrm{Ind}_P^G(\sigma \otimes \nu \otimes 1_N)$ are irreducible. So the set of tempered representations of G can be identified with

$$(\hat{M} \times \hat{A})/W.$$

Example 2.5.37. Let $G = \mathrm{SL}(2,\mathbb{R})$, and $P = MAN$, the cuspidal parabolic subgroup of upper-triangular matrices from Example 2.4.23. It is now possible to prove directly that the algebra of intertwiners of the spherical principal series representation $P_0^+ = \mathrm{Ind}_P^G(1_M \otimes 1_A \otimes 1_N)$ is one-dimensional; see Section II.5 in [20]. By Schur's lemma (Lemma 2.5.2), this implies that P_0^+ is irreducible, and that $W'_{1_M,1_A} = W_{1_M,1_A}$, so $R_{1_M,1_A} = \{e\}$.

The representation space of the nonspherical principal series representation $P_0^- = \mathrm{Ind}_P^G(\sigma \otimes 1_A \otimes 1_N)$ (where $\sigma \in \hat{M}$ is the nontrivial irreducible representation) has two invariant subspaces. These are

- the space V_+ of the restrictions to $\mathbb{R}$ of the continuous extensions of holomorphic functions f on the upper half-plane for which

$$\sup_{y>0} \int_{\mathbb{R}} |f(x+iy)|^2 \, dx < \infty;$$

  and
- the space $V_- = \{\bar{f}; f \in V_+\}$.

See Section II.5 in [20]. We saw in Example 2.5.21 that $W_{\sigma,1_A} = W = \mathbb{Z}/2\mathbb{Z}$, so Corollary 2.5.3 and Theorem 2.5.16 imply that P_0^- has at most two irreducible summands. We find that this representation decomposes into two inequivalent irreducible representations. These are by definition the limits of the discrete series $D_0^{\pm}$ of $\mathrm{SL}(2,\mathbb{R})$.

By Corollary 2.5.34, we now have $R_{\sigma,1_A} = W = \mathbb{Z}/2\mathbb{Z}$ and $W'_{\sigma,1_A} = \{e\}$.

(Note that in this example, we used explicit knowledge of the representations $P^{\pm}_{1_A}$ to determine R-groups.)

Combining the discussion in Subsection 2.3.38, Examples 2.5.21 and 2.5.37, and Theorems 2.4.28 and 2.5.23, we obtain the classification of the tempered representations of $\mathrm{SL}(2,\mathbb{R})$ announced in Example 2.3.30.

Remark 2.5.38. *Knapp and Zuckerman [22, 23] classified all tempered representations of connected, semisimple, linear Lie groups. These are the representations* $\mathrm{Ind}_P^G(\sigma \otimes \nu \otimes 1_N)$, *where* $P = MAN < G$ *is a parabolic subgroup,* σ *is a discrete series or* limit of discrete series *representation of* M *(see Section 1 of [22]), the group* $R_{\sigma,\nu}$ *is trivial, and the representation* $\mathrm{Ind}_P^G(\sigma \otimes \nu \otimes 1_N)$ *is given by* nondegenerate data *(see later (14.127) in [20], or Section 12 of [23]). There are equivalences between these representations described by Weyl group actions. See Theorem 14.91 in [20] or Theorem 14.2 in [23].*

Together with the Langlands classification of admissible representations in terms of tempered representations of subgroups (Theorem 8.54 in [20] or [5, 25]), this completes the classification of admissible representations of connected, semisimple, linear Lie groups.

2.6 Group C^*-Algebras

We still consider a connected, semisimple, linear Lie group G. We end this chapter by using the theory from Sections 2.3, 2.4 and 2.5 to describe the reduced group C^*-algebra of G, and its K-theory. This is based on work by Wassermann [35] and Clare, Crisp, Higson, Song, and Tang [8, 9] (see the final section of Aubert's chapter for another approach).

2.6.1 Group Actions Used In the Computation of Reduced Group C^*-algebras

Consider a cuspidal parabolic subgroup $P = MAN < G$. Let σ be a discrete series representation of M. Via the compact picture of parabolically induced representations (see Subsection 2.4.27), every representation $\mathrm{Ind}_P^G(\sigma \otimes \nu \otimes 1_N)$, for $\nu \in \hat{A}$, can be realized in the same representation space

$$\mathrm{Ind}_P^G(V_\sigma) := \mathrm{Ind}_P^G(V_\sigma \otimes \mathbb{C}_{1_A}).$$

For $\nu \in \hat{A}$ and $w \in W_{\sigma,1_A}$, the operator $\sigma(w)\mathscr{A}_P(w,\sigma,\nu)$ from Corollary 2.5.14 may then be viewed as a unitary operator $U_{w,\nu} \in \mathrm{U}(\mathrm{Ind}_P^G(V_\sigma))$ that intertwines the representations $\mathrm{Ind}_P^G(\sigma \otimes \nu \otimes 1_N)$ and $\mathrm{Ind}_P^G(\sigma \otimes w \cdot \nu \otimes 1_N)$ of G in $\mathrm{Ind}_P^G(V_\sigma)$.

Consider the C^*-algebra $C_0(\hat{A}, \mathcal{K}(\mathrm{Ind}_P^G(V_\sigma)))$, and the group

$$W_\sigma := W_{\sigma, 1_A}. \tag{2.6.1.1}$$

For $w \in W_\sigma$ and $\varphi \in C_0(\hat{A}, \mathcal{K}(\mathrm{Ind}_P^G(V_\sigma)))$, define $w \cdot \varphi \in C_0(\hat{A}, \mathcal{K}(\mathrm{Ind}_P^G(V_\sigma)))$ by

$$(w \cdot \varphi)(\nu) := U_{w, w^{-1}\nu} \varphi(w^{-1}\nu) U^*_{w, w^{-1}\nu}. \tag{2.6.1.2}$$

for all $\nu \in \hat{A}$.

Lemma 2.6.2. *This defines a group action by W_σ on $C_0(\hat{A}, \mathcal{K}(\mathrm{Ind}_P^G(V_\sigma)))$.*

Proof. Let $w, w' \in W_\sigma$. Then for all $\nu \in \hat{A}$,

$$(w \cdot (w' \cdot \varphi))(\nu) = U_{w, w^{-1}\nu} U_{w', w'^{-1}w^{-1}\nu} \varphi(w'^{-1}w^{-1}\nu) U^*_{w', w'^{-1}w^{-1}\nu} U^*_{w, w^{-1}\nu}. \tag{2.6.2.1}$$

By Lemma 2.5.32, there is a $c \in \mathrm{U}(1)$ such that

$$U_{w, w^{-1}\nu} U_{w', w'^{-1}w^{-1}\nu} = c U_{ww', w'^{-1}w^{-1}\nu}$$

So the right-hand side of (2.6.2.1) equals

$$|c|^2 U_{ww', w'^{-1}w^{-1}\nu} \varphi(w'^{-1}w^{-1}\nu) U^*_{ww', w'^{-1}w^{-1}\nu} = ((ww') \cdot \varphi)(\nu).$$

□

Example 2.6.3. Let $G = \mathrm{SL}(2, \mathbb{R})$, and consider the spherical principal series representations $P^+_\nu = \mathrm{Ind}_P^G(1_M \otimes \nu \otimes 1_N)$ from Example 2.4.33, where $1_M \in \hat{M}$ is the trivial representation, and $\nu \in \hat{A} \cong \mathbb{R}$. The representation P^+_ν in $V = L^2(\mathbb{R})$ is irreducible for all $\nu \in \mathbb{R}$, see Example 2.5.37.

We saw in Examples 2.5.11.1 and 2.5.21 that W_{1_M} has two elements, and that the nontrivial element w acts on $\hat{A} \cong \mathbb{R}$ by reflection in the origin. By Schur's lemma, Lemma 2.5.2, irreducibility of P^+_ν implies that the intertwining operator $U_{w,\nu}$ is a multiple of the identity for all $\nu \in \mathbb{R}$. So for all $\nu \in \mathbb{R}$ and $\varphi \in C_0(\hat{A}, \mathcal{K}(\mathrm{Ind}_P^G(V_{1_M})) = C_0(\mathbb{R}, \mathcal{K}(L^2(\mathbb{R})))$,

$$(w \cdot \varphi)(\nu) = \varphi(-\nu).$$

The fixed-point algebra $C_0(\hat{A}, \mathcal{K}(\mathrm{Ind}_P^G(V_{1_M})))^{W_{1_M}}$ for this action is $C_0([0, \infty)) \otimes \mathcal{K}(L^2(\mathbb{R}))$.

Example 2.6.4. Let $G = \mathrm{SL}(2, \mathbb{R})$, and consider the nonspherical principal series representations $P^-_\nu = \mathrm{Ind}_P^G(\sigma \otimes \nu \otimes 1_N)$ from Example 2.4.33, where $\sigma \in \hat{M}$ is the nontrivial irreducible representation, and $\nu \in \hat{A} \cong \mathbb{R}$. The representation P_ν^- in $V = L^2(\mathbb{R})$ is irreducible for all nonzero $\nu \in \mathbb{R}$ and decomposes into two irreducible subspaces $V_\pm \subset L^2(\mathbb{R})$ if $\nu = 0$; see Example 2.5.37. We write $\mathcal{K}(L^2(\mathbb{R})) = M_2(\mathcal{K})$ with respect to the decomposition $L^2(\mathbb{R}) = V_+ \oplus V_-$. (Here $\mathcal{K}$ stands for the algebra of compact operators on any infinite-dimensional, separable Hilbert space.) We claim that the action by the nontrivial element $w \in W_\sigma$ on $C_0(\hat{A}, \mathcal{K}(\mathrm{Ind}_P^G(V_\sigma)) = C_0(\mathbb{R}, M_2(\mathcal{K}))$ satisfies

$$(w \cdot \varphi)(0) = \begin{pmatrix} 1 & 0 \\ 0 & -1 \end{pmatrix} \varphi(0) \begin{pmatrix} 1 & 0 \\ 0 & -1 \end{pmatrix}, \tag{2.6.4.1}$$

for all $\varphi \in C_0(\mathbb{R}, M_2(\mathcal{K}))$.

Because $U_{w,0}$ is a unitary intertwining operator and $L^2(\mathbb{R})$ decomposes into two inequivalent irreducible subspaces, Schur's lemma implies that there are $\lambda_\pm \in \mathrm{U}(1)$ such that, with respect to the decomposition $L^2(\mathbb{R}) = V_+ \oplus V_-$,

$$U_{w,0} = \begin{pmatrix} \lambda_+ & 0 \\ 0 & \lambda_- \end{pmatrix}.$$

The complex conjugation map $C\colon L^2(\mathbb{R}) \to L^2(\mathbb{R})$ interchanges V_+ and V_-, and its square is the identity. The explicit form of the intertwining operators in Example 2.5.7 implies that for all $\nu \in \mathbb{R}$,

$$CU_{w,\nu}C = U_{w,-\nu}.$$

For $\nu = 0$, this implies that $\lambda_\pm$ are real, and hence lie in $\{\pm 1\}$.

By Theorem 2.5.16, $U_{w,0}$ is not a multiple of the identity operator. We conclude that

$$U_{w,0} = \pm \begin{pmatrix} 1 & 0 \\ 0 & -1 \end{pmatrix}.$$

This implies (2.6.4.1).

The fixed-point algebra of the action is now

$$C_0(\hat{A}, \mathcal{K}(\mathrm{Ind}_P^G(V_\sigma)))^{W_\sigma} = \{\varphi \in C_0([0,\infty)) \otimes M_2(\mathcal{K}); \varphi(0) \in \mathcal{K} \oplus \mathcal{K} \subset M_2(\mathcal{K})\}. \tag{2.6.4.2}$$

Analogously to (2.6.1.1), we write

$$\begin{aligned} W'_\sigma &= W'_{\sigma,1_A}; \\ R_\sigma &= R_{\sigma,1_A}. \end{aligned} \tag{2.6.4.3}$$

Let L^{R_σ} be the left regular representation of R_σ in $l^2(R_\sigma)$. This defines an action by R_σ on $C_0(\mathfrak{a}/W'_\sigma, \mathrm{End}(l^2(R_\sigma)))$, by

$$(w \cdot \psi)(X) = L_w^{R_\sigma} \psi(\mathrm{Ad}(w)^{-1}X)(L_w^{R_\sigma})^*, \tag{2.6.4.4}$$

for $w \in R_\sigma$, $\psi \in C_0(\mathfrak{a}/W'_\sigma, \mathrm{End}(l^2(R_\sigma)))$, and $X \in \mathfrak{a}/W'_\sigma$. Here we use the fact that R_σ normalizes W'_σ (point 1. in Theorem 2.5.28), and that $K \cap M$ acts trivially on $\mathfrak{a}$, so the adjoint action indeed descends to an action by R_σ on $\mathfrak{a}/W'_\sigma$.

In the description of the reduced group C^*-algebra of G in Subsection 2.6, the algebras $C_0(\hat{A}, \mathcal{K}(\mathrm{Ind}_P^G(V_\sigma)))^{W_\sigma}$ and $C_0(\mathfrak{a}/W'_\sigma, \mathrm{End}(l^2(R_\sigma))^{R_\sigma}$ of fixed points of the actions (2.6.1.2) and (2.6.4.4), respectively, are used.

Finally, let $A_{\max}$ be the lowest-dimensional factor A that occurs in a cuspidal parabolic subgroup of M. (The subscript max refers to the fact that the corresponding cuspidal parabolic subgroup is maximal.)

Theorem 2.6.5. *Suppose that* $W'_\sigma = \{e\}$. *Then* $R_\sigma = (\mathbb{Z}/2\mathbb{Z})^{\dim(A)-\dim(A_{\max})}$, *and* $\mathfrak{a}$ *is* R_σ*-equivariantly isomorphic to* $\mathbb{R}^{\dim(A)}$, *on which* R_σ *acts by reflections in the first* $\dim(A) - \dim(A_{\max})$ *coordinates.*

See Lemma 12 in [35] or Theorem 3.7 in [9].

Example 2.6.6. Consider the nonspherical principal series of $\mathrm{SL}(2,\mathbb{R})$, as in Example 2.6.4.

We saw in Examples 2.5.11.1 and 2.5.37 that the nontrivial element of $R_\sigma = W_\sigma = W = \mathbb{Z}/2\mathbb{Z}$ acts on $\mathfrak{a} \cong \mathbb{R}$ as reflection in the origin, which is consistent with Theorem 2.6.5. (Now $A_{\max} = \{e\}$.)

Now $R_\sigma = \mathbb{Z}/2\mathbb{Z}$ and $W'_\sigma = \{e\}$. And the nontrivial element $w \in R_\sigma$ acts on $l^2(R_\sigma) = \mathbb{C}^2$ as the matrix $\begin{pmatrix} 0 & 1 \\ 1 & 0 \end{pmatrix}$. So

$$C_0(\mathfrak{a}/W'_\sigma, \mathrm{End}(l^2(R_\sigma))) = C_0(\mathbb{R}, M_2(\mathbb{C})),$$

on which the action (2.6.4.4) is given by

$$(w \cdot \psi)(X) = \begin{pmatrix} 0 & 1 \\ 1 & 0 \end{pmatrix} \psi(-X) \begin{pmatrix} 0 & 1 \\ 1 & 0 \end{pmatrix},$$

for $\psi \in C_0(\mathbb{R}, M_2(\mathbb{C}))$ and $X \in \mathbb{R}$. So by Lemma 2.52 in [37], the fixed-point algebra $C_0(\mathfrak{a}/W'_\sigma, \mathrm{End}(l^2(R_\sigma))^{R_\sigma}$ is isomorphic to the crossed product $C_0(\mathbb{R}) \rtimes \mathbb{Z}/2\mathbb{Z}$.

Example 2.6.7. Suppose that G has discrete series representations, and that $P = G$. Then W is the trivial group, see Example 2.5.6. Let σ be a discrete series representation of $M = G$. Now the group W_σ, and hence its subgroups W'_σ and R_σ is trivial. Furthermore, $\mathfrak{a} = \{0\}$. So the action 2.6.1.2 is the trivial action by the trivial group on $\mathcal{K}(V_\sigma)$. The action (2.6.4.4) is the trivial action by the trivial group on $\mathrm{End}(l^2(R_\sigma))$.

2.6.8 The Reduced Group C^*-Algebra of a Connected, Linear, Semisimple Lie Group

Recall from Theorem 2.3.23 that for every $f \in L^1(G) \cap L^2(G)$, and every $\pi \in \hat{G}$, the operator $\hat{f}(\pi) = \pi(f)$ on V_π is Hilbert–Schmidt, hence compact.

Let $P = MAN < G$ be a cuspidal parabolic subgroup, and σ a discrete series representation of M. Let $C_0(\hat{A}, \mathcal{K}(\mathrm{Ind}_P^G(V_\sigma)))^{W_\sigma}$ be the fixed-point algebra with respect to the action in Lemma 2.6.2.

Proposition 2.6.9. *For $f \in L^1(G) \cap L^2(G)$, define the map $\hat{A} \to \mathcal{K}(\mathrm{Ind}_P^G(V_\sigma))$ by $\nu \mapsto \mathrm{Ind}_P^G(\sigma \otimes \nu \otimes 1_N)(f)$. This construction extends continuously to a $*$-homomorphism from $C_r^*(G)$ to $C_0(\hat{A}, \mathcal{K}(\mathrm{Ind}_P^G(V_\sigma)))^{W_\sigma}$.*

See Corollary 12 in [8] and Remark 2.13 in [9].

As in Theorem 2.5.24, fix θ-stable representatives of conjugacy classes of Cartan subgroups of G, choices of positive restricted roots for each, and resulting cuspidal parabolics $P_j = M_j A_j N_j < G$, for $j = 1, \ldots, n$. For each j, fix representatives σ_j of the orbits of $W_j = N_K(\mathfrak{a}_j)/(K \cap M_j)$ on the set $(\hat{M}_j)_{\mathrm{ds}}$ of discrete series representations of M_j. We write $W_{\sigma_j} := (W_j)_{\sigma_j}$

Theorem 2.6.10. *The $*$-homomorphisms from Proposition 2.6.9 combine to a $*$-isomorphism*

$$C_r^*(G) \to \bigoplus_{j=1}^{n} \bigoplus_{[\sigma_j] \in (\hat{M}_j)_{\mathrm{ds}}/W_j} C_0(\hat{A}_j, \mathcal{K}(\mathrm{Ind}_{P_j}^G(V_{\sigma_j})))^{W_{\sigma_j}}. \qquad (2.6.10.1)$$

See Theorem 6.8 in [8].

Example 2.6.11. Suppose that G has discrete series representations. Then $P = G$ is a cuspidal parabolic subgroup of G. For this parabolic, we have $A = \{e\}$ and $M = G$, so W is trivial. Therefore, every discrete series representation σ of G contributes a term $\mathcal{K}(V_\sigma)$ to the right-hand side of (2.6.10.1).

If $G = K$ is compact, then this implies that

$$C^*(K) \cong \bigoplus_{\sigma \in \hat{K}} \mathcal{K}(V_\sigma).$$

Example 2.6.12. Suppose that G is complex. Then it has one conjugacy class of Cartan subgroups, so in Theorem 2.6.10, only one cuspidal parabolic $P = MAN$ occurs. Now $M = T$ is a maximal torus in K, and $\mathfrak{a} = i\mathfrak{t}$. So $W = N_K(\mathfrak{a})/(K \cap M) = N_K(T)/T =: W_K$ is the Weyl group of (K,T). Because T is compact, all its irreducible representations belong to the discrete series. And for all $\sigma \in \hat{T}$, we have $R_\sigma = \{e\}$ by Theorem 2.5.35, so $W'_\sigma = W_\sigma =$

$(W_K)_\sigma$. So (2.6.10.1) becomes

$$C_r^*(G) \cong \bigoplus_{[\sigma]\in\hat{T}/W_K} C_0(\mathfrak{t}^*/(W_K)_\sigma)\otimes\mathcal{K} = C_0((\hat{T}\times\mathfrak{t}^*)/W_K)\otimes\mathcal{K}.$$

This is Proposition 4.1 in [31].

Example 2.6.13. Let $G = \mathrm{SL}(2,\mathbb{R})$. By Example 2.6.11 and Section 2.3.38, the right-hand side of (2.6.10.1) contains

$$\bigoplus_{n=1}^{\infty} \mathcal{K}(V_{D_n^+}) \oplus \mathcal{K}(V_{D_n^-}).$$

Consider the upper-triangular parabolic P from Example 2.4.23, associated with the noncompact Cartan subgroup H. By Example 2.6.3, the trivial representation of $M = \{\pm I\}$ contributes a term $C^\infty([0,\infty))\otimes\mathcal{K}$ to the right-hand side of (2.6.10.1). By Example 2.6.4, the nontrivial representation of M contributes a term (2.6.4.2). So by Theorem 2.6.10,

$$\begin{aligned} C_r^*(\mathrm{SL}(2,\mathbb{R})) \cong \bigoplus_{n=1}^{\infty} &\mathcal{K}(V_{D_n^+}) \oplus \mathcal{K}(V_{D_n^+}) \oplus (C^\infty([0,\infty))\otimes\mathcal{K}) \\ &\oplus \{\varphi \in C_0([0,\infty))\otimes M_2(\mathcal{K}); \varphi(0)\in\mathcal{K}\oplus\mathcal{K}\}. \end{aligned}$$

2.6.14 K-Theory

We now turn to a simpler description of reduced group C^*-algebras, up to Morita equivalence, which is suitable for computing K-theory. For a cuspidal parabolic $P = MAN < G$, and a discrete series representation σ of M, consider the decomposition $W_\sigma = W'_\sigma \rtimes R_\sigma$ from Theorem 2.5.28. The group R_σ acts on

$$C_0(\hat{A}/W'_\sigma)\otimes\mathcal{K}(\mathrm{Ind}_P^G(V_\sigma), l^2(R_\sigma))$$

via the operators $U_{w,\nu}$ on $\mathrm{Ind}_P^G(V_\sigma)$ (for $w\in R_\sigma$ and $\nu\in\hat{A}$), and the left regular action by R_σ on $l^2(R_\sigma)$. The fixed-point space

$$(C_0(\hat{A}/W'_\sigma)\otimes\mathcal{K}(\mathrm{Ind}_P^G(V_\sigma), l^2(R_\sigma)))^{R_\sigma} \tag{2.6.14.1}$$

of this action is a right Hilbert module over

$$(C_0(\hat{A}/W'_\sigma)\otimes\mathcal{K}(\mathrm{Ind}_P^G(V_\sigma))^{R_\sigma} = (C_0(\hat{A})\otimes\mathcal{K}(\mathrm{Ind}_P^G(V_\sigma))^{W_\sigma},$$

and a left module over $(C_0(\hat{A}/W'_\sigma)\otimes\mathcal{K}(l^2(R_\sigma))^{R_\sigma}$.

Theorem 2.6.15. *The module* (2.6.14.1) *implements a Morita-equivalence between* $(C_0(\hat{A})\otimes\mathcal{K}(\mathrm{Ind}_P^G(V_\sigma))^{W_\sigma}$ *and* $(C_0(\hat{A}/W'_\sigma)\otimes\mathcal{K}(l^2(R_\sigma))^{R_\sigma}$.

See Corollary 7 in [35] and Theorem 2.25 in [9].

The algebra $(C_0(\hat{A}/W'_\sigma) \otimes \mathcal{K}(l^2(R_\sigma))^{R_\sigma}$ can alternatively be described as a crossed product.

Lemma 2.6.16. *There is a unique* $*$*-isomorphism from* $C_0(\hat{A}/W'_\sigma) \rtimes R_\sigma$ *to* $(C_0(\hat{A}/W'_\sigma) \otimes \mathcal{K}(l^2(R_\sigma))^{R_\sigma}$ *that sends* $\varphi \in C_0(\hat{A}/W'_\sigma)$ *to* $\sum_{r\in R} r \cdot \varphi \otimes e_r$*, with* $e_r \in \mathcal{K}(l^2(R_\sigma)$ *the rank-one projection onto functions supported at* r*, and* $r \in R_\sigma$ *to the right-regular action by* r *on* $l^2(R_\sigma)$*.*

See Proposition 4.3 in [32].

We write $W_{\sigma_j} = W'_{\sigma_j} \rtimes R_{\sigma_j}$ as in Theorem 2.5.28.

Theorem 2.6.17. *The isomorphism in Theorem 2.6.10 and the Morita equivalences in Theorem 2.6.15 combine to form a Morita equivalence*

$$C_r^*(G) \sim \bigoplus_{j=1}^{n} \bigoplus_{[\sigma_j]\in(\hat{M}_j)_{\mathrm{ds}}/W_j} C_0(\hat{A}_j/W'_{\sigma_j}, \mathcal{K}(l^2(R_{\sigma_j})))^{R_{\sigma_j}}. \qquad (2.6.17.1)$$

See Theorem 8 in [35] and Theorem 2.31 in [9].

Example 2.6.18. If G has discrete series representations, so $P = G$ is a cuspidal parabolic subgroup, then R_σ is trivial for all discrete series representations σ of G. So each discrete series representation contributes a term $\mathbb{C}$ to the right-hand side of (2.6.17.1).

Example 2.6.19. If G is complex, then as in Example 2.6.12, we find that (2.6.17.1) becomes

$$C_r^*(G) \sim \bigoplus_{[\sigma]\in\hat{T}/W_K} C_0(\mathfrak{t}^*/(W_K)_\sigma) \cong C_0((\hat{T}\times\mathfrak{t}^*)/W_K).$$

Example 2.6.20. Let $G = \mathrm{SL}(2,\mathbb{R})$, and let $P = MAN$ be the cuspidal parabolic from Example 2.4.23 associated with the noncompact Cartan subgroup H.

For the trivial representation 1_M of M, we have $W'_{1_M} = W_{1_M} = \mathbb{Z}/2\mathbb{Z}$, and $R_{1_M} = \{e\}$. So this representation contributes a term $C_0([0,\infty))$ to the right-hand side of (2.6.17.1).

By Example 2.6.6, the nontrivial representation of M contributes a term

$$\left\{\psi \in C_0(\mathbb{R}, M_2(\mathbb{C})); \psi(X) = \begin{pmatrix}0 & 1\\ 1 & 0\end{pmatrix}\psi(-X)\begin{pmatrix}0 & 1\\ 1 & 0\end{pmatrix} \text{ for all } X\in\mathbb{R}\right\}$$
$$\cong C_0(\mathbb{R}) \rtimes (\mathbb{Z}/2\mathbb{Z})$$

to the right-hand side of (2.6.17.1).

Together with Example 2.6.18, we obtain a Morita equivalence

$$C_r^*(\mathrm{SL}(2,\mathbb{R})) \sim \bigoplus_{n=1}^{\infty}(\mathbb{C}\oplus\mathbb{C}) \oplus C_0([0,\infty)) \oplus C_0(\mathbb{R}) \rtimes (\mathbb{Z}/2\mathbb{Z}). \qquad (2.6.20.1)$$

This is Example 4.8 in [6].

Theorem 2.6.17 can be used to compute the K-theory of $C_r^*(G)$.

Lemma 2.6.21. *If $P = MAN < G$ is a cuspidal parabolic subgroup and σ is a discrete series representation of M such that $W'_\sigma \neq \{e\}$, then*

$$K_d(C_0(\hat{A}/W'_\sigma, \mathcal{K}(l^2(R_\sigma)))^{R_\sigma}) = 0,$$

for $d = 0$ and $d = 1$.

See Theorem 4.2 in [9].

Lemma 2.6.22. *If $P = MAN < G$ is a cuspidal parabolic subgroup and σ is a discrete series representation of M such that $W'_\sigma = \{e\}$, then*

$$K_{\dim(G/K)}(C_0(\hat{A}/W'_\sigma, \mathcal{K}(l^2(R_\sigma)))^{R_\sigma}) = \mathbb{Z}, \qquad (2.6.22.1)$$

and

$$K_{\dim(G/K)+1}(C_0(\hat{A}/W'_\sigma, \mathcal{K}(l^2(R_\sigma)))^{R_\sigma}) = 0.$$

See Theorem 4.3 in [9]. This is based on Theorem 2.6.5.

In the setting of Lemma 2.6.22, we choose a generator $b(P,\sigma)$ of (2.6.22.1). Theorem 2.6.17 and Lemmas 2.6.21 and 2.6.22 have the following consequence.

Corollary 2.6.23. *We have $K_{\dim(G/K)+1}(C_r^*(G)) = 0$, and*

$$K_{\dim(G/K)}(C_r^*(G)) = \bigoplus_{j=1}^{n} \bigoplus_{[\sigma_j]\in(\hat{M}_j)_{\mathrm{ds}}/W_j;W'_{\sigma_j}=\{e\}} \mathbb{Z}b(P_j,\sigma_j).$$

Example 2.6.24. If G has discrete series representations, then each discrete series representation σ contributes a term $\mathbb{Z}b(G,\sigma)$ to $K_0(C_r^*(G))$. This was already clear from Example 2.6.11.

Example 2.6.25. If G is complex, then Corollary 2.6.23 becomes

$$K_{\dim(G/K)}(C_r^*(G)) = \bigoplus_{[\sigma]\in\hat{T}/W_K;(W_K)_\sigma=\{e\}} \mathbb{Z}b(P,\sigma),$$

while $K_{\dim(G/K)+1}(C_r^*(G)) = 0$. (Now $\dim(G/K) = \dim(K) \equiv \dim(T)$ modulo 2.) This is consistent with Examples 2.6.12 and 2.6.19, because $C_0(\mathfrak{t}^*/(W_K)_\sigma)$

has trivial K-theory if $(W_K)_\sigma$ is nontrivial. (This follows by a simpler version of the proof of Theorem 4.2 in [9].)

Example 2.6.26. In the case of $G = \mathrm{SL}(2,\mathbb{R})$, let $P = MAN$ be the upper-triangular cuspidal parabolic, and σ the nontrivial representation of M. Corollary 2.6.23 states that

$$K_0(C_r^*(\mathrm{SL}(2,\mathbb{R}))) = \left(\bigoplus_{n=1}^{\infty} \mathbb{Z}b(G, D_n^+) \oplus \mathbb{Z}b(G, D_n^-)\right) \oplus \mathbb{Z}b(P, \sigma),$$

and $K_1(C_r^*(\mathrm{SL}(2,\mathbb{R}))) = 0$. Note that the term $C_0([0,\infty))$ in (2.6.20.1) does not contribute to K-theory.

References

[1] J. Adams and O. Taïbi. *Galois and Cartan cohomology of real groups*. Duke Math. J. 167 (6) (2018), 1057–1097.

[2] J. Adams, M. van Leeuwen, P. Trapa, and D. A. Vogan, Jr. *Unitary representations of real reductive groups*. arXiv:1212.2192, 2012.

[3] J. Adams, M. van Leeuwen, P. Trapa, D. A. Vogan, Jr., et al. *Atlas software package*. www.liegroups.org.

[4] J. Adams and D. A. Vogan, Jr. *L-groups, projective representations, and the Langlands classification*. Amer. J. Math. 114 (1) (1992), 45–138.

[5] E. P. van den Ban. Induced representations and the Langlands classification. In *Representation theory and automorphic forms* (Edinburgh, 1996), volume 61 of Proceedings of Symposia in Pure Mathematics, pages 123–155. American Mathematical Society, Providence, RI, 1997.

[6] P. Baum, A. Connes, and N. Higson. Classifying space for proper actions and K-theory of group C^*-algebras. In *C^*-algebras*: 1943–1993 (San Antonio, TX, 1993), volume 167 of Contemporary Mathematics, pages 240–291. American Mathematical Society, Providence, RI, 1994.

[7] E. Cartan. *Les groupes projectifs qui ne laissent invariante aucune multiplicité plane*. Bull. Soc. Math. France, 41 (1913), 53–96.

[8] P. Clare, T. Crisp, and N. Higson. *Parabolic induction and restriction via C^*-algebras and Hilbert C^*-modules*. Compos. Math. 152 (6) (2016), 1286–1318.

[9] P. Clare, N. Higson, Y. Song, and X. Tang. *On the Connes-Kasparov isomorphism, I: The reduced C^*-algebra of a real reductive group and the K-theory of the tempered dual*. Jpn. J. Math. 19 (1) (2024), 67–109.

[10] M. Cowling, U. Haagerup, and R. Howe. *Almost L^2 matrix coefficients*. J. Reine Angew. Math. 387 (1988), 97–110.

[11] J. Dixmier. C^*-algebras. In: *North-Holland mathematical library*, volume 15. North-Holland Publishing Co., Amsterdam-New York-Oxford, 1977. Translated from the French by Francis Jellett.

[12] Harish-Chandra. *Representations of a semisimple Lie group on a Banach space. I.* Trans. Amer. Math. Soc. 75 (1953), 185–243.

[13] Harish-Chandra. *Discrete series for semisimple Lie groups. II. Explicit determination of the characters.* Acta Math. 116 (1966), 1–111.

[14] Harish-Chandra. *Harmonic analysis on real reductive groups. I. The theory of the constant term.* J. Funct. Anal. 19 (1975), 104–204.

[15] Harish-Chandra. *Harmonic analysis on real reductive groups. II. Wavepackets in the Schwartz space.* Invent. Math. 36 (1976), 1–55.

[16] Harish-Chandra. *Harmonic analysis on real reductive groups. III. The Maass-Selberg relations and the Plancherel formula.* Ann. Math. 104 (1) (1976), 117–201.

[17] E. Kaniuth and K. F. Taylor. *Induced representations of locally compact groups,* volume 197, Cambridge Tracts in Mathematics. Cambridge University Press, Cambridge, 2013.

[18] A. W. Knapp and E. M. Stein. *Intertwining operators for semisimple groups. II.* Invent. Math. 60 (1) (1980), 9–84.

[19] A. W. Knapp. *Commutativity of intertwining operators for semisimple groups.* Compositio Math. 46 (1) (1982), 33–84.

[20] A. W. Knapp. *Representation theory of semisimple groups.* An overview based on examples. Princeton Mathematical Series, 36. Princeton University Press, Princeton, NJ, 1986.

[21] A. W. Knapp. *Lie groups beyond an introduction*, 2nd edition. Progress in Mathematics, 140. Birkhäuser Boston, Inc., Boston, MA, 2002.

[22] A. W. Knapp and G. J. Zuckerman. *Classification of irreducible tempered representations of semisimple groups.* Ann. Math. 116 (2) (1982), 389–455.

[23] A. W. Knapp and G. J. Zuckerman. *Classification of irreducible tempered representations of semisimple groups. II.* Ann. Math. 116 (3) (1982), 457–501.

[24] P. Koosis. *An irreducible unitary representation of a compact group is finite dimensional.* Proc. Amer. Math. Soc. 8 (1957), 712–715.

[25] R. P. Langlands. On the classification of irreducible representations of real algebraic groups. In *Representation theory and harmonic analysis on semisimple Lie groups*, volume 31 of Mathematical Surveys and Monographs, pages 101–170. American Mathematical Society, Providence, RI, 1989.

[26] T. Matsuki. *The orbits of affine symmetric spaces under the action of minimal parabolic subgroups.* J. Math. Soc. Japan. 31 (2) (1979), 331–357.

[27] F. I. Mautner. *Unitary representations of locally compact groups. I.* Ann. Math. 51 (2) (1950), 1–25.

[28] F. I. Mautner. *Unitary representations of locally compact groups. II.* Ann. Math. 52 (2) (1950), 528–556.

[29] D. Miličić. *The dual spaces of almost connected reductive groups.* Glasnik Mat. Ser. III. 9 (29) (1974), 273–288.

[30] L. Nachbin. *On the finite dimensionality of every irreducible unitary representation of a compact group.* Proc. Amer. Math. Soc. 12 (1961), 11–12.

[31] M. G. Penington and R. J. Plymen. *The Dirac operator and the principal series for complex semisimple Lie groups.* J. Funct. Anal. 53 (3) (1983), 269–286.

[32] M. A. Rieffel. *Actions of finite groups on C^*-algebras.* Math. Scand., 47(1) (1980), 157–176.

[33] I. E. Segal. *An extension of Plancherel's formula to separable unimodular groups*. Ann. Math. 52 (2) (1950), 272–292.

[34] W. Forrest Stinespring. *A semi-simple matrix group is of type I*. Proc. Amer. Math. Soc. 9 (1958), 965–967.

[35] A. Wassermann. *Une démonstration de la conjecture de Connes-Kasparov pour les groupes de Lie linéaires connexes réductifs*. C. R. Acad. Sci. Paris Sér. I Math. 304 (18) (1987), 559–562.

[36] H. Weyl. *Theorie der Darstellung kontinuierlicher halb-einfacher Gruppen durch lineare Transformationen. I*. Math. Z. 23 (1) (1925), 271–309.

[37] D. Williams. *Crossed products of C^*-algebras*, volume 134 of Mathematical Surveys and Monographs. American Mathematical Society, Providence, RI, 2007.

3

Dirac Operators and Representation Theory

Hang Wang

3.1 Introduction

This chapter is an introduction to the geometric perspective of tempered unitary representations of connected reductive Lie groups via Dirac induction and the Connes–Kasparov isomorphism. Dirac operators were initially introduced from physics when Dirac was looking for first-order differential operators whose square would equal the Laplacian in Lorentzian geometry. They were then adapted to Riemannian geometry and became an effective tool in different areas of mathematics. For example, in geometry, a classical Dirac operator on a spin manifold M satisfies the Lichnerowich formula

$$D^2 = \nabla^*\nabla + \frac{\kappa}{4}$$

where κ is the scalar curvature of M. The formula suggests deep relations between geometry and Dirac operators and has been useful in problems of finding metrics of positive scalar curvature on M and their obstruction. In this chapter, we shall see that Dirac operators associated with a noncompact semisimple or reductive Lie group G are keys in understanding links between geometry, representation theory, and operator algebras. In fact, the central objects of this chapter are Dirac-type operators on homogeneous spaces of the form G/K where K is a maximal compact subgroup of G. Such operators are geometric models for components of the tempered representations of G having nontrivial K-theory. Furthermore, the parametrization of the Dirac-type operators by irreducible unitary representations of K has a one-to-one correspondence with these components. The correspondence is intimately related to the Lichnerowich-type formula in the context of representation theory and eventually leads to the Connes–Kasparov isomorphism. The Connes–Kasparov isomorphism is a verified case of the Baum–Connes conjecture, which is an important open problem in noncommutative geometry [8, 9]. There are many ways of showing the Connes–Kasparov isomorphism, but the highlight of these notes is to present a proof initiated by Penington and Plymen using harmonic analysis and the Fourier transform for Dirac-type operators on G/K [44]. For simplicity, we shall also focus on the class of complex semisimple Lie groups they worked on.

The structure of this chapter is the following. In Section 3.2, we review formulations of Dirac operators. In Section 3.3, we recall some classical Dirac operators on homogeneous spaces arising in representation theory. In Section 3.4, we recall the theory of higher index, a language setting, in order to formulate the Connes–Kasparov conjecture. In Section 3.5 we introduce the Dirac induction, the Connes–Kasparov conjecture, and survey a couple of ways for proving it. Section 3.6 is devoted to the harmonic-analytic approach to the Connes–Kasparov isomorphism using the example of connected semisimple groups.

This chapter serves as extended notes for the lectures at the London Mathematical Society Bath Symposium "Operator K-Theory and Representation Theory" held at the University of Bath, UK, during the dates 19–23 July 2021. The notes are likely not comprehensive due to the limited vision of the author. However, it is hoped that these could develop some curiosities into this research direction and promote new findings in relevant areas. I would like to thank the organizers, Nigel Higson, Roger Plymen, and Haluk Şengün, for this opportunity of creating the notes.

3.2 Dirac Operators

In this section, we overview the definition of Dirac operators. According to Quillen, Dirac operators are quantization of the theory of connections[12]. This means that a Dirac operator could be the Levi–Civita connection $\nabla : C^\infty(M) \to C^\infty(M \otimes T^*M)$ on a Riemannian manifold M followed by some quantization map $c : TM \to \mathrm{End}(S)$ for some vector bundle S over M. Here, TM and T^*M are identified through the Riemannian metric. Then, it would be reasonable to assume the existence of a bundle S viewed as a module over TM, which carries a connection ∇^S compatible with the one on TM, and to define the Dirac operator following the compositions

$$C^\infty(S) \xrightarrow{\nabla^S} C^\infty(S \otimes T^*M) \xrightarrow{\cong} C^\infty(S \otimes TM) \xrightarrow{c} C^\infty(S).$$

Since the Dirac operators considered in this chapter are on homogeneous spaces, we shall reduce the engagement of differential geometry by giving an axiomatic definition for Dirac-type operators before the classical definition.

3.2.1 Dirac-Type Operators

Dirac-type operators form a class of operators reflecting the geometric features of the manifold being acted on. Let M be a Riemannian manifold and $S \to M$ a Hermitian vector bundle. Denote by $C_c^\infty(M, S)$ the vector space of compactly supported smooth sections of S.

Definition 3.2.2. A *Dirac-type operator*

$$D : C_c^\infty(M,S) \to C_c^\infty(M,S)$$

is a symmetric first-order differential operator satisfying

$$[D, f]^2 = -\|df\|^2 I. \tag{3.2.2.1}$$

Here, I stands for the identity operator.

Remark 3.2.3. *The relation (3.2.2.1) can be used to recover the metric from the operator D in the context of the more general setting of a spectral triple. See [21, VI.1]. For the canonical Dirac operator D associated with a spin^c structure of M, the equality in (3.2.2.1) is closely related to the defining relation for Clifford algebras:*

$$c(df)^2 = -\|df\|^2 I,$$

where the identification $[D, f] = c(\mathrm{d}f)$ motivates the replacement of $\mathrm{d}f$ *by the commutator $[D, f]$ in quantized calculus in the theory of spectral triples. See [21, IV.1].*

Some classical examples of Dirac-type operators are presented below.

Example 3.2.4. Let $M = \mathbb{R}^3$ and $S = M \times \mathbb{C}^2 \to M$ be the trivial bundle. Set

$$D = \sum_{i=1}^{3} \sigma_i \frac{\partial}{\partial x_i} : C_c^\infty(\mathbb{R}^3, S) \to C_c^\infty(\mathbb{R}^3, S)$$

where

$$\sigma_1 = \begin{bmatrix} i & 0 \\ 0 & -i \end{bmatrix}, \quad \sigma_2 = \begin{bmatrix} 0 & -1 \\ 1 & 0 \end{bmatrix}, \quad \sigma_3 = \begin{bmatrix} 0 & i \\ i & 0 \end{bmatrix}$$

are Pauli matrices. Note that

$$\sigma_i \sigma_j + \sigma_j \sigma_i = -2\delta_{ij} \qquad \forall i, j \tag{3.2.4.1}$$

To see this is a Dirac-type operator, let us check (3.2.2.1). For $s \in C_c(\mathbb{R}^3, \mathbb{R}^3 \times \mathbb{C}^2)$ and $f \in C_c^\infty(\mathbb{R}^2)$,

$$[D, f]s = \sum_{i=1}^{3} \sigma_i \frac{\partial}{\partial x_i}(fs) - \sum_{i=1}^{3} f \sigma_i \frac{\partial}{\partial x_i} s = \sum_{i=1}^{3} \sigma_i \left(\frac{\partial}{\partial x_i} f \right) s.$$

Then using (3.2.4.1), one has

$$[D, f]^2 = \sum_{i,j=1}^{3} \sigma_i \sigma_j f_{x_i} f_{x_j} = -\|df\|^2.$$

The equality (3.2.4.1) can be generalized to the defining relation of a Clifford algebra. Let $\{e_i\}_{1 \le i \le n}$ be an orthonormal basis for $\mathbb{R}^n$. We will define an algebra generated by these elements. Let $V = \mathbb{R}^n$ and $T(V) = \mathbb{R} \oplus V \oplus (V \otimes V) \oplus (V \otimes V \otimes V) \oplus \cdots$ be the tensor algebra. Let I be the ideal generated by $e_i \otimes e_j + e_j \otimes e_i + 2\delta_{ij}, \forall 1 \le i, j \le n$. Denote by $Cl(\mathbb{R}^n)$ the quotient $T(V)/I$ and by $c(e_i)$ the element in $Cl(\mathbb{R}^n)$ corresponding to e_i.

Definition 3.2.5. The *Clifford algebra* $Cl_n := Cl(\mathbb{R}^n)$ is the algebra generated by

$$c_i := c(e_i), \qquad 1 \le i \le n$$

subject to relations

$$c_i c_j + c_j c_i = -2\delta_{ij} \qquad \forall 1 \le i, j \le n.$$

Let M be an oriented Riemannian manifold of dimension n. Then one can form the oriented orthonormal frame bundle P_{SO} over M, which is known as the principal $SO(n)$-bundle of TM, that is, $P_{SO} \times_{SO(n)} \mathbb{R}^n \cong TM$. The algebra Cl_n is $\mathbb{Z}_2$-graded:

$$Cl_n = (Cl_n)_0 \oplus (Cl_n)_1$$

where $(Cl_n)_0$ is the subalgebra generated by $c_i \cdot c_j$ for all $1 \le i, j \le n$, and $(Cl_n)_1$ is the vector space generated by $c_{i_1} \cdot c_{i_2} \cdots c_{i_k}$ with k odd. The *Clifford algebra* of M is the associated bundle

$$Cl(TM) := P_{SO} \times_{SO(n)} Cl_n.$$

Here, the special orthogonal group $SO(n)$ acts on $\mathbb{R}^n$ by rotations through matrix multiplication and the action extends to an action on Cl_n.

It can be verified that for a non-negative integer k

$$Cl_{2k} \cong M_{2^k}(\mathbb{C}) \qquad Cl_{2k+1} \cong M_{2^k}(\mathbb{C}) \oplus M_{2^k}(\mathbb{C}). \tag{3.2.5.1}$$

See [40, 45] for example. Under this identification, $e_i \mapsto c(e_i)$ can be regarded as irreducible representations of Cl_n on $\mathbb{C}^{2^k}$. Thus, we also denote the above isomorphism by c:

$$c : Cl_{2k} \to M_{2^k}(\mathbb{C}) \qquad c : Cl_{2k+1} \to M_{2^k}(\mathbb{C}) \oplus M_{2^k}(\mathbb{C}).$$

Example 3.2.6. On $\mathbb{R}^n$, there is a Dirac-type operator D given by

$$D = \sum_{i=1}^{n} c_i \frac{\partial}{\partial x_i} : C_c^\infty(\mathbb{R}^n, S) \to C_c^\infty(\mathbb{R}^n, S)$$

satisfying $D^2 = -\frac{\partial^2}{\partial x_1^2} - \cdots - \frac{\partial^2}{\partial x_n^2}$. Here, $S = \mathbb{R}^n \times \mathbb{C}^{2^k} \to \mathbb{R}^n$, where $k = \lfloor \frac{\dim M}{2} \rfloor$, and $\mathbb{C}^{2^k}$ is the irreducible representation of Cl_n determined by the isomorphisms (3.2.5.1). When n is even, S is $\mathbb{Z}_2$ graded, that is, $S = S^+ \oplus S^-$ so that $c(v_i)S^\pm \subset S^\mp$. With respect to this grading, D has the form

$$D = \begin{bmatrix} 0 & D^- \\ D^+ & 0 \end{bmatrix} \qquad (D^+)^* = D^-$$

and is referred to as an odd operator.

Example 3.2.7. De Rham operators, signature operators, and Dolbeault operators are all Dirac-type operators. For more details, see, for example, [12, 45].

- Let M be an oriented Riemannian manifold and $S = \Lambda^* M$ be the exterior algebra bundle over M. The $\mathbb{Z}_2$-grading of S is determined by the parity of differential forms:

$$S^+ = \Lambda^{even} M \qquad S^- = \Lambda^{odd} M.$$

  The de Rham operator $D = d + d^*$ on M is an odd symmetric operator of the form

$$D = \begin{bmatrix} 0 & d + d^*|_{\Lambda^{odd}} \\ d + d^*|_{\Lambda^{even}} & 0 \end{bmatrix} : C_c^\infty(M, S^+ \oplus S^-) \to C_c^\infty(M, S^+ \oplus S^-).$$

- Let M be a Riemannian manifold of dimension divisible by 4. The Hodge $*$-operator satisfies $*^2 = I$. With respect to the grading given by the ± 1 eigenspaces $\Lambda^\pm$ of $*$, the de Rham differential is an odd symmetric operator

$$D = \begin{bmatrix} 0 & d + d^*|_{\Lambda^-} \\ d + d^*|_{\Lambda^+} & 0 \end{bmatrix} : C_c^\infty(M, \Lambda^+ \oplus \Lambda^-) \to C_c^\infty(M, \Lambda^+ \oplus \Lambda^-),$$

  called the signature operator.
- Let M be a complex manifold or a Riemannian manifold with an almost complex structure. Let $\Omega^{0,even}(M)$ and $\Omega^{0,odd}(M)$ be even and odd antiholomorphic forms on M, respectively. The de Rham differential decomposes into $d = \partial + \bar{\partial}$ where $\partial : \Omega^{*,*}(M) \to \Omega^{*+1,*}(M)$ and $\bar{\partial} : \Omega^{*,*}(M) \to \Omega^{*,*+1}(M)$. The Dolbeault operator is given by

$\bar{\partial} + \bar{\partial}^* : \Omega^{0,*}(M) \to \Omega^{0,*}(M)$, which is an odd symmetric operator of the form

$$\begin{bmatrix} 0 & \bar{\partial} + \bar{\partial}^*|_{\Omega^{0,odd}} \\ \bar{\partial} + \bar{\partial}^*|_{\Omega^{0,even}} & 0 \end{bmatrix} : \Omega^{0,even}(M) \oplus \Omega^{0,odd}(M)$$
$$\to \Omega^{0,even}(M) \oplus \Omega^{0,odd}(M).$$

Example 3.2.8. One way of creating Dirac-type operators is by twisting a Dirac-type operator by vector bundles. Given a Dirac-type operator $D : C_c^\infty(M,S) \to C_c^\infty(M,S)$ and a vector bundle $V \to M$, by Swan's theorem, there exists a vector bundle $W \to M$ such that $V \oplus W \cong M \times \mathbb{C}^N$. Thus

$$C_c^\infty(M, S \otimes V) \oplus C_c^\infty(M, S \otimes W) \cong C_c^\infty(M,S)^N.$$

Let

$$P : C_c^\infty(M, S \otimes (M \times \mathbb{C}^N)) \to C_c^\infty(M, S \otimes (M \times \mathbb{C}^N))$$

be the operator determined by the projection from the free $C_c^\infty(M)$-module $C_c^\infty(M, M \times \mathbb{C}^N)$ to $C_c^\infty(M,V)$ as a projective $C_c^\infty(M)$-module. The operator

$$D_V : C_c^\infty(M, S \otimes V) \to C_c^\infty(M, S \otimes V)$$
$$D_V := P \begin{bmatrix} D \otimes 1 & & \\ & \ddots & \\ & & D \otimes 1 \end{bmatrix}_{N \times N}$$

is a Dirac-type operator, known as a twisted Dirac operator.

Remark 3.2.9.
- *Any Dirac-type operator is locally a "minimal" Dirac operator twisted by a vector bundle. In fact,* $2^{\lfloor \frac{\dim M}{2} \rfloor} | \dim S$. *For* $\mathbb{R}^n$, *it is easy to check that in Example 3.2.6, the dimension of* S *is* $2^{\lfloor \frac{\dim M}{2} \rfloor}$ *and is minimal so that the Dirac-type operator can be defined.*
- *If the dimension of* S *is minimal, that is,* $\dim S = 2^{\lfloor \frac{\dim M}{2} \rfloor}$, *then* D *is called a canonical Dirac operator on* M.
- *In general, such a bundle* S *does not always exist globally, except that* M *satisfies a topological condition, known as a* Spin^c *structure.*

3.2.10 Spin^c-Structure and the Canonical Dirac Operator

Recall that $\pi_1(SO(n)) = \mathbb{Z}_2$ when $n \geq 3$ and $\pi_1(SO(2)) = \mathbb{Z}$. Let the spin group Spin(n) be the double cover of $SO(n)$. Thus, there is an extension of groups:

$$1 \to \mathbb{Z}_2 \to \text{Spin}(n) \to SO(n) \to 1.$$

The spin group Spin(n) can be identified as a subgroup of Cl_n. In fact, the multiplicative subgroup of Cl_k generated by unit vectors of the Euclidean space $\mathbb{R}^n$ is denoted by $\mathrm{Pin}(n)$. Then $\mathrm{Spin}(n)$ is the intersection of $\mathrm{Pin}(n)$ and the even part of $(Cl_n)_0$ of the Clifford algebra. The double cover structure is given by the adjoint action of $\mathrm{Spin}(n)$ on $\mathbb{R}^n$:

$$\xi_0 : \mathrm{Spin}(n) \to SO(\mathbb{R}^n) \qquad v \mapsto (w \mapsto vwv^{-1}).$$

The irreducible representations of Spin(n) are closely related to those of Cl_n. See [45].

Define the spin^c-group $\mathrm{Spin}^c(n)$ to be the balanced product $\mathrm{Spin}(n) \times_{\mathbb{Z}_2} U(1)$.

Definition 3.2.11. An oriented Riemannian manifold M admits a *Spin structure* (or a *Spin^c structure*) if the transition functions $\rho_{\alpha\beta} : U_\alpha \cap U_\beta \to SO(n)$ admit a lift

$$\tilde{\rho}_{\alpha\beta} : U_\alpha \cap U_\beta \to \mathrm{Spin(n)(or\ Spin}^c\mathrm{(n))}$$

satisfying the cocycle condition $\tilde{\rho}_{\alpha\beta}\tilde{\rho}_{\beta\gamma}\tilde{\rho}_{\gamma\alpha} = 1$ so that the tangent bundle can be equipped with the structure group Spin(n) (or Spin^c(n)). Equivalently, let P_{SO} be the principal $SO(n)$-bundle over M, then M has a *Spin structure* if P_{SO} admits a double cover, namely a principal Spin(n)-bundle

$$\xi : P_{\mathrm{Spin}} \to P_{SO}$$

such that

$$\xi(pg) = \xi(p)\xi_0(g) \qquad \forall p \in P_{\mathrm{Spin}},\ \forall g \in \mathrm{Spin}(n).$$

Remark 3.2.12. *1. If M admits a Spin structure, then $TM \cong P_{SO} \times_{SO(n)} \mathbb{R}^n$ can be equivalently expressed as*

$$TM \cong P_{\mathrm{Spin}} \times_{\mathrm{Spin}(n)} \mathbb{R}^n.$$

2. *If M is a complex manifold, then M has a Spin^c-structure.*
3. *Spin^c is a condition stronger than orientability and is known as the K-orientation, that is, when M has Spin^c structure, one has*

$$K^*(M) = K_{n-*}(M)$$

the K-theoretic Poincaré duality. Here, the left-hand side is the topological K-theory of M and the right-hand side is the K-homology of M. See [53] and the reference therein for more details.

Definition 3.2.13. Let M be a Riemannian manifold. A vector bundle $S \to M$ is a *Clifford module* if

- it is a fiberwise module over the Clifford algebra $Cl(TM)$ of the tangent bundle and
- it admits a Hermitian metric $\langle\cdot,\cdot\rangle$ and a connection ∇^S compatible with the Levi–Civita connection ∇^{TM}:

$$\langle c(v)s_1, s_2\rangle + \langle s_1, c(v)s_2\rangle = 0 \qquad v \in TM, s_i \in C_c^\infty(M,S)$$
$$\nabla_X^S(Ys) = (\nabla_X^{TM}Y)s + Y\nabla_X^S s \qquad X,Y \in C_c^\infty(M,TM), s \in C_c^\infty(M,S).$$

Let M be a Riemannian manifold with a Spin^c structure. A *spinor bundle* $S \to M$ is a Clifford module where each fiber is the irreducible representation of the corresponding fiber of the Clifford algebra $Cl(TM)$. Denote by Δ the irreducible representation of Cl_n where n is the dimension of M. Then the spinor bundle can be explicitly written as

$$S = P_{\mathrm{Spin}(n)} \times_{\mathrm{Spin}(n)} \Delta.$$

Here, Spin(n)$\subset Cl_n$ acts on Δ via Clifford multiplications.

Example 3.2.14. Let M be an oriented Riemannian manifold. Then the complexified exterior algebra $\Lambda_{\mathbb{C}}^* M := \Lambda^* M \otimes \mathbb{C}$ is a Clifford module [40].

If M has a Spin^c structure, then there exists a global vector bundle $S \to M$ so that we have vector bundle isomorphisms:

$$\Lambda_{\mathbb{C}}^* M \cong S \otimes S^* \cong \mathrm{End}(S).$$

Here, S is the spinor bundle over M. There also exists a canonical Dirac operator

$$D : C_c^\infty(M,S) \to C_c^\infty(M,S)$$

such that $d + d^* = D_{S^*}$.

In particular, if M has an almost complex structure (this implies that M has a spin^c-structure), then the antiholomorphic forms, that is, sections of $\Lambda^{0,*}(M) \to M$ is a model for S and the Dolbeault operator $\bar{\partial} + \bar{\partial}^*$ is the canonical Dirac operator.

Proposition 3.2.15 (See for example [12]). *If M is an even-dimensional manifold with a Spin^c structure, then there is a spinor bundle $S \to M$ with* $\dim S = 2^{\lfloor \frac{\dim M}{2} \rfloor}$ *so that the Clifford multiplication of $e \in TM$ on S gives rise to an isomorphism*

$$c : Cl(TM) \to \mathrm{End}(S).$$

Definition 3.2.16. Let M be a Riemannian manifold with a Spin^c structure and S be the associated spinor bundle. The *(canonical) Dirac operator* is given by

$$D = \sum_{i=1}^{n} c(e_i)\nabla^S_{e_i} : C_c^\infty(M,S) \to C_c^\infty(M,S)$$

where $\{e_i\}$ is a local orthonormal frame for TM and ∇^S is a connection on S compatible with the Levi–Civita connection.

The spin^c connection ∇^S can be constructed whenever M has a Spin^c structure. Explicit constructions can be found in [40].

Remark 3.2.17. *1. The above-defined Dirac operator D is a formally self-adjoint, first-order differential operator satisfying*

$$c(df)^2 = -\|df\|^2.$$

It is a Dirac-type operator with the minimal possible dimension of the Clifford module it acts on.

2. If M has even dimension, then $S = S^+ \oplus S^-$ is $\mathbb{Z}_2$-graded and D has the form

$$D = \begin{bmatrix} 0 & D^- \\ D^+ & 0 \end{bmatrix} \qquad (D^+)^* = D^-.$$

3.3 Dirac Operators in Representation Theory

Dirac-type operators appear to be an important geometric tool in the study of representation theory, which dates back to the birth of Atiyah–Singer index theory. Bott [13] investigated a class of Dirac-type operators on homogeneous vector bundles over spaces of the form G/H, where G is a compact connected Lie group and H is a closed connected subgroup. He introduced a refined index taken values in the character ring of the group G. This refined index is then used to produce representations of compact Lie groups coming from geometry via Dirac-type operators. Then, following Harish-Chandra's seminal work on the classification of discrete series representations of semisimple Lie groups G [26, 27], Parthasarathy and Atiyah–Schmid used L^2-kernels of twisted Dirac operators on G/K to construct discrete series of G, where K is a maximal compact subgroup of G [5, 43]. Vogan introduced an algebraic construction of representations of semisimple Lie groups by introducing Dirac cohomology [32]. Dirac operators arise as Dirac elements in Connes–Kasparov isomorphism, a formula of computing the K-theory of the reduced group C^*-algebra of G where G is an almost connected group. The isomorphism is a

C^*-algebraic analogue of Mackey correspondence by the well-known work of Higson [29]. The Connes–Kasparov conjecture has been verified using various methods in different contexts by Pennington–Plymen [44], Wassermann [51], Lafforgue [39], Chabert–Echterhoff–Nest [16], Clare–Crisp–Higson [17], and many others.

3.3.1 Representation Ring and Characters of Compact Lie Groups

Let G be a locally compact group. Recall that a *unitary representation* (π, H_π) is a continuous homomorphism

$$\pi : G \to U(H_\pi)$$

where H_π is a Hilbert space and $U(H_\pi)$ is the group of unitary operators on H_π. The representation π is *irreducible* if H_π has no nontrivial G-invariant closed subspaces. Denote by $\widehat{G}$ the set of unitary equivalence classes of irreducible unitary representations of G, also known as the *unitary dual* of G. The unitary dual $\widehat{G}$ admits the hull kernel topology, defined via primitive ideals of the maximal group C^*-algebra of G, making it a topological space [24]. Furthermore, it has a Plancherel measure whose support is known as the tempered dual of G and is denoted by $\widehat{G}_t$. In this subsection, we only focus on compact Lie groups. Let G be a compact Lie group. It is well known that every irreducible representation π of G has finite dimension.

Example 3.3.2. The *one*-dimensional irreducible unitary representations of the unit circle group $G = S^1$ are given by

$$\pi_n : S^1 \to GL_1(\mathbb{C}) = \mathbb{C}\backslash\{0\} \qquad z \to z^n$$

for every $n \in \mathbb{Z}$. It is known that all irreducible representations of a locally compact abelian group are *one*-dimensional.

Denote by $R(G)$ the *representation ring* of G. It consists of "linear combinations" ($\mathbb{Z}$-coefficient) of irreducible representations, or alternatively, formal difference of finite dimensional representations:

$$R(G) := \{[V] - [W] : V, W \text{ finite dimensional representation of } G\}. \quad (3.3.2.1)$$

Here, $[V]$ stands for the unitary equivalence class of the representation V. The additive and multiplicative structures are given respectively by the direct sum and tensor of two representations.

Given a representation π of G, one can produce its *character* χ_π:

$$\chi_\pi(g) := \mathrm{Tr}[\pi(g) : H_\pi \to H_\pi] \qquad g \in G.$$

It follows from the trace property that χ is constant on each conjugacy class and that $\chi_\pi = \chi_{\pi'}$ whenever π and π' are unitary equivalent. The character value can be explicitly calculated by the Weyl character formula. Similar to the representation ring, the set of characters forms a ring under addition and pointwise multiplication and is compatible with the operations in $R(G)$:

$$\chi_\pi + \chi_\rho = \chi_{\pi\oplus\rho} \qquad \chi_\pi \cdot \chi_\rho = \chi_{\pi\otimes\rho}.$$

Conversely, any smooth function on G that is invariant on each conjugacy class is the difference of the characters of two representations of G. It is a classical fact in representation theory that $R(G)$ can be identified with the ring of characters.

Example 3.3.3. Consider $G = SO(3)$ with the maximal torus $SO(2)$ identified with the unit circle S^1 (or denoted by $\mathbb{T}^1$). Let $(\pi_n, V_n) \in \widehat{SO(3)}$ be labeled by its highest weight n, that is,

$$V_n|_{S^1} \cong \bigoplus_{j=0}^{2n} \mathbb{C}_{j-n}$$

where $\mathbb{C}_j$ stands for $\mathbb{C}$ on which S^1 acts by $g \cdot z = g^j z, g \in S^1, z \in \mathbb{C}$. Then

$$\chi_{\pi_n}(g) = \sum_{j=0}^{2n} g^{j-n} \qquad g \in S^1.$$

Rewriting $g \in S^1$ by $\begin{bmatrix} \cos\theta & \sin\theta \\ -\sin\theta & \cos\theta \end{bmatrix}$, then

$$\chi_{\pi_n}\left(\begin{bmatrix} \cos\theta & \sin\theta \\ -\sin\theta & \cos\theta \end{bmatrix}\right) = \frac{e^{i(n+\frac{1}{2})\theta} - e^{-i(n+\frac{1}{2})\theta}}{e^{i\frac{\theta}{2}} - e^{-i\frac{\theta}{2}}} = \frac{\sin(n+\frac{1}{2})\theta}{\sin\frac{\theta}{2}}.$$

3.3.4 Equivariant Index

Let G be a compact Lie group acting on a closed Riemannian manifold M preserving the metric. Let $R(G)$ be the representation ring of G as in (3.3.2.1) and identified as rings of characters. Let D be a Dirac-type operator on M. Because the local expression of Dirac operator is independent of the orthonormal basis, the isometric action of G on M commutes with D:

$$Dg = gD \qquad \forall g \in G.$$

Such property will be referred to as "G-invariance." Dirac operators on M are typical examples of *G-invariant elliptic operators*. We refer to the definition of ellipticity in [40].

Definition 3.3.5. The *equivariant index* of a G-invariant elliptic operator $D = \begin{bmatrix} 0 & D^- \\ D^+ & 0 \end{bmatrix}$ on M, where $(D^+)^* = D^-$ is given by

$$\mathrm{ind}_G D = [\ker D^+] - [\ker D^-] \in R(G);$$

It is determined by the characters

$$\mathrm{ind}_G D(g) := \mathrm{Tr}(g|_{\ker D^+}) - \mathrm{Tr}(g|_{\ker D^-}) \in \mathbb{C} \qquad \forall g \in G.$$

The equivariant index is a notion that generalizes the Fredholm index of D. Note that the Fredholm index of D is

$$\dim \ker D^+ - \dim \ker D^- = \mathrm{Tr}(e|_{\ker D^+}) - \mathrm{Tr}(e|_{\ker D^-})$$

which is the equivariant index at the group identity e.

Example 3.3.6. Consider the de Rham operator on a closed, oriented even-dimensional manifold M:

$$D^+ = d + d^* : \Omega^{even}(M) \to \Omega^{odd}(M)$$
$$D^- = d + d^* : \Omega^{odd}(M) \to \Omega^{even}(M).$$

Recall that $\ker D$ is the space of harmonic forms on M and by Hodge theorem, $\ker D^+$ and $\ker D^-$ are identified with $H^{even}_{DR}(M,\mathbb{R})$ and $H^{odd}_{DR}(M,\mathbb{R})$, respectively. The equivariant index of the de Rham operator is known as the Lefschetz number associated with the isometry $g \in G$:

$$\begin{aligned} L(g) := \mathrm{ind}_G D(g) &= \mathrm{Tr}(g|_{\ker D^+}) - \mathrm{Tr}(g|_{\ker D^-}) \\ &= \sum_{i \geq 0} (-1)^i \mathrm{Tr}\left[g_{*,i} : H_i(M,\mathbb{R}) \to H_i(M,\mathbb{R})\right]. \end{aligned}$$

If $L(g) \neq 0$, then g has a fixed point in M. In fact, following the Atiyah–Segal–Singer index theorem, there is a cohomological formula for the equivariant index localized at the fixed point submanifold $M^g = \{x \in M | gx = x\}$. See [6].

3.3.7 Equivariant Index and Representations

The geometric construction of representations of a compact Lie group via equivariant indices dates back to Bott when he studied homogenous vector bundles [14]. If G is a connected complex semisimple Lie group and P a parabolic subgroup, then G is a principal P bundle over the homogeneous manifold $M = G/P$. A holomorphic representation $\pi : P \to \mathrm{Aut}(V)$ on a complex vector space V induces a holomorphic vector bundle given by the balanced product

$$E := G \times_P V := (G \times V)/\sim$$

where $(gp, v) \sim (g, \phi(p)v)$ for all $g \in G, p \in P$, and $v \in V$. The homogeneous vector bundle E admits the left G action:

$$k(g, v) = (kg, v) \qquad \forall k, g \in G, v \in V.$$

Thus, the cohomology $H^k(M, \mathcal{O}(E))$ with coefficients in the holomorphic sections of E is a G-module. In [14], Bott showed that if π is irreducible, then $H^k(M, \mathcal{O}(E))$ vanishes except possibly for some degree q, and $H^q(M, \mathcal{O}(E))$ is an irreducible representation of G whose highest weight is associated with π. The theorem generalizes a theorem of Borel and Weil and is a geometric realization of induced representations.

Observe that a holomorphic section can be identified as an element of the kernel of the associated Dolbeault operator. Bott introduced the equivariant index for invariant elliptic operators on homogeneous spaces, independent of Atiyah and Singer's work, and produced examples where analytic issues can be covered by representation theory. See [13].

Let G be a connected semisimple Lie group and T be a maximal torus. Recall that an irreducible representation of G is labeled by its highest weight in $\widehat{T}$. The homogeneous space $M = G/T$ is a G-space and admits a complex structure. Thus, one has the Dolbeault operator $\bar{\partial}$ on M. For $\pi \in \widehat{G}$, let $\lambda \in \widehat{T}$ be its highest weight. Similarly to the above holomorphic vector bundle, we construct the complex line bundle

$$L_\lambda := G \times_T \mathbb{C}_\lambda.$$

Then, by Borel–Weil–Bott, the geometric realization of representations can be explicitly described as in the following:

Theorem 3.3.8 (Borel–Weil–Bott). *The character of an irreducible representation π of a compact group G is given by the equivariant index of the twisted Dolbeault operator*

$$\bar{\partial}_{L_\lambda} + \bar{\partial}^*_{L_\lambda}$$

on the homogenous space G/T.

Equivalently, by the Atiyah–Bott fixed point formula [4], the equivariant index can be calculated purely in terms of the information of π:

Theorem 3.3.9 (Atiyah–Bott). *Let π be an irreducible representation of a compact group G with the highest weight λ. Then, for a regular point g in T,*

$$\operatorname{ind}_G(\bar{\partial}_{L_\lambda} + \bar{\partial}^*_{L_\lambda})(g) = \frac{\sum_{w \in W} \det(w) e^{w(\lambda+\rho)}}{e^\rho \prod_{\alpha \in \Delta^+}(1 - e^{-\alpha})}(g)$$

where the right-hand side is the Weyl character formula for π. Here W stands for the Weyl group and ρ is the half sum of the positive root for (G,T).

Example 3.3.10. Let $\bar{\partial}_n + \bar{\partial}_n^*$ be the Dolbeault–Dirac operator on

$$S^2 \cong SO(3)/\mathbb{T}^1,$$

coupled to the line bundle

$$L_n := SO(3) \times_{\mathbb{T}^1} \mathbb{C}_n \to S^2.$$

By Borel–Weil–Bott, the equivariant index is

$$\mathrm{ind}_{SO(3)}(\bar{\partial}_n + \bar{\partial}_n^*) = [V_n] \in R(SO(3)).$$

By the Atiyah–Segal–Singer's formula or the Atiyah–Bott fixed point formula, one recovers the Weyl character formula:

$$\mathrm{ind}_{SO(3)}(\bar{\partial}_n + \bar{\partial}_n^*)(g) = \frac{g^n}{1-g^{-1}} + \frac{g^{-n}}{1-g} = \sum_{j=0}^{2n} g^{j-n}.$$

Compare Example 3.3.3.

3.3.11 Dirac Operators on Homogeneous Spaces

In preparation to generalize the equivariant index theory to noncompact Lie groups, we introduce Dirac operators on symmetric spaces of noncompact type. Refer to [43] for more details.

Let us first recall the so-called equivariant spin structure on homogeneous spaces. Let G be a connected noncompact semisimple Lie group and $\mathfrak{g}$ be the Lie algebra of G, identified as the left invariant vector fields on G. Let K be a maximal compact subgroup of G and $\mathfrak{k}$ its Lie algebra. Denote by G/K the homogeneous space made up of the left cosets of K in G. The Killing form B on the complexification of $\mathfrak{g}$ gives rise to the Cartan decomposition

$$\mathfrak{g} = \mathfrak{k} + \mathfrak{p}$$

where $\mathfrak{p}$ is defined by $\{Y \in \mathfrak{g} | B(X,Y) = 0, \forall X \in \mathfrak{k}\}$. The space $\mathfrak{p}$ can be identified with

$$V := T_{x_0}(G/K),$$

the tangent space of G/K at $x_0 := eK \in G/K$. Because the restriction of the Killing form B on $\mathfrak{p}$ is positive definite, it induces a K-invariant inner product on $\mathfrak{p}$. In fact, denoting $SO(\mathfrak{p})$ as the rotation group of $\mathfrak{p}$ with respect to the Killing form, the adjoint action of K on $\mathfrak{p}$ gives rise to a homomorphism

$$Ad : K \to SO(\mathfrak{p}).$$

This induces a G-invariant Riemannian metric on the manifold G/K.

Definition 3.3.12. The Riemannian symmetric space $M = G/K$ has a *G-equivariant spin structure* if $Ad : K \to SO(\mathfrak{p})$ can be lifted to

$$\tilde{Ad} : K \to \mathrm{Spin}(\mathfrak{p})$$

so that $\pi \circ \tilde{Ad} = Ad$, where $\pi : \mathrm{Spin}(\mathfrak{p}) \to SO(\mathfrak{p})$ is the double cover map.

Remark 3.3.13. *The principal bundle over G/K of orthonormal frames can be identified as $F := G \times_{K,Ad} SO(\mathfrak{p})$. Given a G-equivariant spin structure, one can construct*

$$\tilde{F} := G \times_{K,\tilde{Ad}} \mathrm{Spin}(\mathfrak{p}), \qquad (3.3.13.1)$$

the principal $Spin(\mathfrak{p})$ bundle over G/K so that $\tilde{F} \times_{Spin(\mathfrak{p})} SO(\mathfrak{p})$ is $SO(\mathfrak{p})$-equivalent to F. Thus, G/K has a spin structure. A G-equivariant spin structure has the additional structure where $F, \tilde{F}$ admit the left actions of G and the canonical map $\tilde{F} \to F, (g,s) \mapsto (g, \pi(s))$ commutes with the action of G.

Remark 3.3.14. *A G-equivariant spin structure can be assumed when G can be replaced by its double cover. See [22].*

Example 3.3.15. Let $G = SL(2,\mathbb{R})$ and $K = SO(2)$. Then the homogeneous space

$$G/K \cong \{z \in \mathbb{C} | \mathrm{Im}(z) > 0\}$$

is a hyperbolic manifold $\mathbb{H}$ with constant negative scalar curvature. In fact, G acts on $\mathbb{H}$ transitively by the Mobius transform: for $g = \begin{bmatrix} a & b \\ c & d \end{bmatrix}$ and $z \in \mathbb{H}$, $gz := \frac{az+b}{cz+d}$. Because the stabilizer of i is isomorphic to $SO(2)$, the homogeneous space G/K is identified with $\mathbb{H}$. Let $\tilde{G}, \tilde{K}$ be the double cover of G, K, then $M = G/K = \tilde{G}/\tilde{K}$ has a $\tilde{G}$-equivariant spin structure.

Next, we introduce Dirac-type operators on the homogeneous space G/K. Assume that G/K admits a G-equivariant spin structure. Denote by Δ the complex spin representation of $\mathrm{Spin}(\mathfrak{p})$. The dimension of Δ is $2^{\dim \mathfrak{p}/2}$. Let $\tilde{F}$ be the principal Spin structure defined in (3.3.13.1). Then the homogeneous vector bundle $S := \tilde{F} \times_{\mathrm{Spin}(n)} \Delta$ is the Spinor bundle over G/K and the smooth sections $C_c^\infty(G/K, S)$ of S with compact support can be identified with $(C_c^\infty(G) \otimes \Delta)^K$, the K-invariant part of the tensor.

Let $Cl(\mathfrak{p})$ be the Clifford algebra of $(\mathfrak{p}, B)$ and the adjoint action of K on $\mathfrak{p}$ extends to the Clifford algebra. When $\mathfrak{p}$ has even dimension, one has an isomorphism

$$c : Cl(\mathfrak{p}) \to \mathrm{End}(\Delta).$$

Compare Proposition 3.2.15. In fact, this isomorphism gives rise to

$$Cl(G/K) \cong G \times_K Cl(\mathfrak{p}) \cong G \times_K \mathrm{End}(\Delta) \cong \mathrm{End}(S).$$

Therefore, the spin Dirac operator in Definition 3.2.16 specializing in homogeneous spaces can be restated as follows:

Definition 3.3.16. Let $\{p_i\}_{i=1}^n$ be an orthonormal basis for $\mathfrak{p}$ with respect to the Killing form B. The *Dirac operator* $D : (C_c^\infty(G) \otimes \Delta)^K \to (C_c^\infty(G) \otimes \Delta)^K$ is given by

$$D = \sum_{i=1}^{n} c(p_i) p_i,$$

where $(Pf)(g) := \frac{d}{dt}|_{t=0} f(e^{-tP} g)$ for $P \in \mathfrak{p}, g \in G$, and $f \in C_c^\infty(G)$. The Dirac operator on the spinor bundle $S = \tilde{F} \times_{Spin(n)} \Delta$ over G/K can be identified as D.

Remark 3.3.17. *The Dirac operator above is G-equivariant by definition, that is, $Dg = gD$ for all $g \in G$.*

Twisted Dirac operators can be defined similarly. Let (π, V) be a finite dimensional representation of K. Then $E := G \times_K V$ is the associated homogeneous vector bundle over G/K, then the canonical Dirac operator on G/K associated with a G-equivariant spin structure twisted by E can be alternatively defined as

$$D_V : (C_c^\infty(G) \otimes \Delta \otimes V)^K \to (C_c^\infty(G) \otimes \Delta \otimes V)^K$$

$$D_V = \sum_{i=1}^{n} c(p_i) p_i \otimes 1 \qquad \{p_i\} \text{ is an orthonormal basis of } \mathfrak{p}.$$

Here, we do not distinguish the notation D_E from D_V when it is clear from the context.

Lichnerowicz theorem calculates the square of the canonical Dirac operator and the relation to the scalar curvature has been of tremendous value to the geometry of Dirac operators. Analogously, in the equivariant setting, there is a formula of the square of D in the language of Lie algebras.

Let G be a connected noncompact semisimple Lie group and K a maximal compact subgroup. Let

$$\Omega_G = -\sum_{j=1}^{\dim K} X_j^2 + \sum_{i=1}^{\dim(G/K)} Y_i^2$$

be the Casimir operator on G, where X_j and Y_i are orthonormal bases for $\mathfrak{k}$ and $\mathfrak{p}$ respectively. This plays the role of Laplacian in the Lichnerowicz theorem. Let T be the maximal torus of K and $\mathfrak{t}$ be its Lie algebra. The Killing form B induces a positive definite scalar product on $\langle \cdot, \cdot \rangle$ on $i\mathfrak{t}^*$. The Weyl group acts on $i\mathfrak{t}^*$ and we choose a closed Weyl chamber C. Let $V = V_\tau$ be an irreducible unitary representation of K with the highest weight $\tau \in C \subset i\mathfrak{t}^*$.

Theorem 3.3.18 ([43, Proposition 3.2]). *Let $D_V : (C_c^\infty(G) \otimes \Delta \otimes V)^K \to (C_c^\infty(G) \otimes \Delta \otimes V)^K$ be the Dirac operator on G/K twisted by V with highest weight τ. Then*

$$D_V^2 = -\Omega_G + \langle \tau + 2\rho_K, \tau \rangle - \langle \rho, \rho \rangle + \langle \rho_K, \rho_K \rangle$$

where ρ_K is the half sum of positive roots of K with respect to T and ρ is the half sum of positive roots of G with respect to T.

This formula is useful in the proof of the Connes–Kasparov isomorphism via the method of the noncommutative Fourier transform.

3.4 Higher Index

Representations for a compact Lie group can be studied geometrically through the equivariant indices of elliptic operators invariant under the group action. Similarly, for a noncomapct group G, a geometric elliptic operator can also be constructed on a manifold, where the manifold admits a proper, cocompact action from the group. For example, the homogeneous space G/K, where K is a maximal compact subgroup, is a proper G-space with one orbit. Unlike the compact setting, twisted Dirac operators on such spaces are no longer Fredholm. To generalize the notion of an equivariant index, it is necessary to regard the operator as a "generalized Fredholm operator" having an index in the K-theory of the C^*-algebra associated with the group. We shall recall two well-known formulations of the higher index via equivariant Roe algebras and via Hilbert C^*-modules, then introduce some concrete descriptions to be used in later sections.

3.4.1 Fredholm Index as a K-Theory Element

Let M be a complete Riemannian manifold and D a Dirac-type operator on M. Then D is essentially self-adjoint. Because the principal symbol $\sigma_D = ic(\xi)$ is invertible when $\xi \neq 0$, the Dirac operator D is elliptic. Assume in addition that M is a closed (compact and without boundary) and D has the form $\begin{bmatrix} 0 & D^- \\ D^+ & 0 \end{bmatrix}$. For example, M is compact spin and D is the canonical Dirac operator on the $\mathbb{Z}_2$-graded spinor bundle S. Then D is Fredholm and the Fredholm index of D is defined by

$$\mathrm{ind} D := \dim\ker D^+ - \dim\ker D^- \in \mathbb{Z}.$$

Since D is an essentially self-adjoint first-order elliptic differential operator on the closed manifold M, by the spectral theory, D has a real discrete spectrum whose absolute values tend to infinity. Also, every element in the spectrum is an eigenvalue of D with finite multiplicity. Then, for a real function $f \in C_0(\mathbb{R})$, the operator $f(D)$ has bounded real spectrum tending to 0, where the nonzero sepctrum are eigenvalues of finite multiplicity. Then, by the spectral theorem again, $f(D)$ is a compact self-adjoint operator. Thus, applying functional calculus, one has a map

$$C_0(\mathbb{R}) \to \mathcal{K}(L^2(M,S)) \qquad f \mapsto f(D)$$

which extends to their multipliers

$$C_b(\mathbb{R}) \to \mathcal{B}(L^2(M,S)) \qquad f \mapsto f(D).$$

Here, C_b stands for the algebra of bounded continuous functions.

One can then normalize D to a bounded Fredholm operator by choosing a continuous odd increasing function

$$\chi : \mathbb{R} \to (-1,1) \text{ such that } \lim_{x\to\infty} \chi(x) = 1. \tag{3.4.1.1}$$

Then, because $\chi^2 - 1 \in C_0(\mathbb{R})$ and

$$\chi(D) \in \mathcal{B}(L^2(M,S)) \quad \text{and} \quad \chi(D)^2 - I \in \mathcal{K}(L^2(M,S)),$$

the normalized operator $\chi(D)$ is Fredholm. Moreover, by the spectral mapping theorem, $\chi(D)$ has the same Fredholm index as D. Recall that a bounded Fredholm operator F is invertible modulo compact operators, hence F gives rise to an element of $K_1(\mathcal{B}/\mathcal{K})$. Then the K-theoretic index can be defined by composing the boundary map ∂ of the K-theoretic six-term exact sequence associated with the short exact sequence:

$$0 \to \mathcal{K}(L^2(M,S)) \to \mathcal{B}(L^2(M,S)) \to \mathcal{B}(L^2(M,S))/\mathcal{K}(L^2(M,S)) \to 0.$$

Definition 3.4.2. The *K-theoretic index* is given by

$$\operatorname{ind}(D) = \partial[\chi(D)] \in K_0(\mathcal{K})$$

where

$$\partial : K_1(\mathcal{B}(L^2(M,S))/\mathcal{K}(L^2(M,S))) \to K_0(\mathcal{K}(L^2(M,S))) \cong \mathbb{Z}$$

is one of the boundary maps.

Remark 3.4.3. *It can be easily verified that different choices of normalizing functions are compact perturbations of one another. See, for example, Theorem 10.6.5 in [30]. This means that the K-theoretic index and the Fredholm index are independent of the choice of χ. In particular, one can choose*

$$\chi(x) = \frac{x}{\sqrt{x^2+1}}.$$

Proposition 3.4.4. *The Fredholm index of D defined by* $\dim\ker D^+ - \dim \ker D^-$ *is equal to the K-theoretic index $\partial[\chi(D)]$. The equality is identified under the isomorphism induced by the trace map* $\mathrm{Tr}_* : K_0(\mathcal{K}) \to \mathbb{Z}$.

Proof. Without loss of generality, we may choose $\chi(x) = \frac{x}{\sqrt{1+x^2}}$ and replace D by $D_t = \sqrt{t}D$. Then

$$\chi(D_t) = \frac{D_t}{\sqrt{1+D_t^2}} = \begin{pmatrix} 0 & a_t^* \\ a_t & 0 \end{pmatrix}$$

where $a_t = (1 + D_t^+D_t^-)^{-\frac{1}{2}}D_t^+$ and $a_t^* = D_t^-(1 + D_t^+D_t^-)^{-\frac{1}{2}}$. By the standard definition of the boundary map in the K-theoretic six-term exact sequence, we obtain:

$$\partial(\chi(D_t)) := \left[\begin{pmatrix} a_ta_t^* & a_t(1-a_t^*a_t)^{\frac{1}{2}} \\ a_t^*(1-a_ta_t^*)^{\frac{1}{2}} & 1-a_t^*a_t \end{pmatrix}\right] - \left[\begin{pmatrix} 1 & 0 \\ 0 & 0 \end{pmatrix}\right] \in K_0(\mathcal{K}).$$

See, for example, Exercise 8.3.D in [52]. Taking trace and setting $g = \frac{1}{1+x}$ and $f = 1 - \chi^2 = \frac{1}{x^2+1}$, we have

$$\mathrm{Tr}(\partial[\chi(D_t)]) = \mathrm{Tr}[(1-a_t^*a_t) - (1-a_ta_t^*)] = \mathrm{Tr}g(tD^-D^+) - \mathrm{Tr}g(tD^+D^-).$$

Note that the above equality is independent of t. Then let $t \to \infty$ and use the fact that D^+D^-, D^-D^+ has discrete and identical spectrum except for 0, we have

$$\begin{aligned}\lim_{t\to\infty}[\mathrm{Tr}g(tD^-D^+) - \mathrm{Tr}g(tD^+D^-)] &= \lim_{t\to\infty} \mathrm{Tr}_s f(\sqrt{t}D) \\ &= \dim(\ker D^+) - \dim(\ker D^-).\end{aligned}$$

Here, Tr_s stands for the supertrace. (Note that one can always replace χ by a nicer function so that $f(\sqrt{t}D)$ is a trace class operator, for example, $f(x) = e^{-x^2}, g = e^{-|x|}$.) The proposition is then proved. □

Remark 3.4.5. *Here we can assume M to be spin even-dimensional and also assume that D is the canonical Dirac operator, which is odd with respect to some $\mathbb{Z}_2$-grading. For the canonical Dirac operator, the odd-dimensional case can be defined similarly using the exponential map*

$$\delta : K_0(\mathscr{B}(L^2(M,S))/\mathscr{K}(L^2(M,S))) \to K_1(\mathscr{K}(L^2(M,S))) \cong 0,$$

which may be nonvanishing for generalized Fredholm operators.

3.4.6 Higher Index via Equivariant Roe Algebras

Let G be a locally compact group. Let M be a complete Riemannian manifold equipped with a proper, cocompact, and isometric action of G. Let D be a Dirac-type operator on $E \to M$. It is G-invariant and essentially self-adjoint because G acts by isometries and M is complete.

Similarly as earlier, choosing a normalizing function χ as in (3.4.1.1), we have $\chi(D) \in \mathscr{B}(L^2(M,S))$. This is no longer Fredholm because $\chi(D)^2 - 1$ may not be a compact operator. However, we could broaden the algebra of compact operators to a suitable algebra so that an index could make sense in the K-theory of this algebra.

Let $H = L^2(G) \otimes L^2(M)$ on which $C_0(M)$ acts by the pointwise multiplication ϕ on $L^2(M)$ and G acts unitarily on H through U, such that the action U is compatible with the action on M, that is,

$$U_g^{-1}\phi(f)U_g = \phi(g^*f) \qquad [g^*(f)](x) := f(g^{-1}x).$$

Definition 3.4.7. A bounded operator $T : H \to H$ is *locally compact* if $T\phi(f)$ and $\phi(f)T$ are compact for all $f \in C_0(M)$. T is said to have *finite propagation* if there exists $R > 0$, such that the Schwartz kernel K_T of T satisfies $K_T(x,y) = 0$ whenever $d(x,y) \geq R$. A bounded operator is *pseudo-local* if $[T,\phi(f)]$ is compact for all $f \in C_0(M)$.

Definition 3.4.8 ([30, Sections 12.3 and 12.5]). The *equivariant Roe algebra $C^*(M)^G$* is defined as the operator norm closure of G-equivariant, locally compact operators with finite propagation. The multiplier algebra $D^*(M)^*$ of $C^*(M)^G$ can be identified with the operator norm closure of G-equivariant, pseudolocal operators with finite propagation.

Remark 3.4.9. *Roe algebras are defined in general for metric spaces, where the so-called ample representations in abstract Hilbert spaces are needed. That means the Hilbert spaces should be large enough.*

Let D be a Dirac-type operator on M. Suppose the Fourier transform of χ is compactly supported, then applying the Fourier transform $\chi(x) = \frac{1}{2\pi} \int_{\mathbb{R}} \hat{\chi}(t)e^{itx}dt$ to D, using functional calculus

$$\chi(D) = \frac{1}{2\pi}\int_{\mathbb{R}} \hat{\chi}(t)e^{itD}dt,$$

one can show that $\chi(D)$ is pseudolocal and has finite propagation, and that $1 - \chi(D)^2$ is locally compact with finite propagation. Then one can show that for any normalizing function χ satisfying (3.4.1.1), one has

$$\chi(D) \in D^*(M)^G \qquad 1 - \chi(D)^2 \in C^*(M)^G. \tag{3.4.9.1}$$

See, for example, Lemma 10.5.5 and Proposition 10.5.6 of [30] and Section 8.2 of [53]. In particular, choose χ so that $1 - \chi^2(x) = e^{-x^2}$, then $e^{-tD^2} \in C^*(M)^G$ for $t > 0$. In fact, the (equivariant) Roe algebra was invented to incorporate features of the heat operator associated with a Dirac-type operator.

Because the action is cocompact, the equivariant Roe algebra is closely related to the reduced group C^*-algebra. Recall that the reduced group C^*-algebra $C^*_r(G)$ is defined to be the operator norm closure of the image of the convolution map

$$C_c(G) \to \mathcal{B}(L^2(G)) \qquad f \mapsto [g \mapsto f * g].$$

The spectrum of this C^*-algebra is known as the tempered dual $\widehat{G}_t$ of the group G. Recall that $\pi \in \widehat{G}_t$ if and only if π is an irreducible unitary representation of G whose matrix coefficient belongs to $L^{2+\epsilon}(G)$ for all $\epsilon > 0$. When G is abelian, there is an isomorphism

$$C^*_r(G) \to C_0(\widehat{G}_t)$$

determined by the Fourier or the Gelfand transform

$$f \mapsto \hat{f} \qquad \hat{f}(\pi) = \int_G f(g)\pi(g)dg.$$

Note that the tempered dual $\widehat{G}_t$ in the case of the abelian group is simply the Pontryagin dual $\mathrm{Hom}(G, U(1))$ of G, and also equal to the full unitary dual $\widehat{G}$ of G due to amenability. When G is a general real reductive group, $C^*_r(G)$ is calculated in the beautiful paper by Clare, Crisp, and Higson [17].

For the following result, see [30, Lemma 12.5.3] when the group is discrete and see [25, Theorem 2.11] for the general case of locally compact groups.

Proposition 3.4.10. *The equivariant Roe algebra is Morita equivalent to $C_r^*(G)$ and hence they have the same K-theory:*

$$K_*(C^*(M)^G) \cong K_*(C_r^*(G)).$$

The proposition ensures the formulation of the higher index of D. In fact, from (3.4.9.1) one knows that

$$[\chi(D)] \in K_1(D^*(M)^G/C^*(M)^G)$$

when D is odd with respect to a $\mathbb{Z}_2$-grading and

$$\left[\frac{\chi(D)+1}{2}\right] \in K_0(D^*(M)^G/C^*(M)^G)$$

when D is ungraded.

Definition 3.4.11 ([30, Sections 10.5–10.6, 12.5])**.** Let D be a Dirac-type operator on M equipped with a proper, cocompact, isometric action by a locally compact group G. The *higher index* of D is defined by

$$\mathrm{ind}_G(D) := \partial_0[\chi(D)] \in K_0(C^*(M)^G) \cong K_0(C_r^*(G))$$

when D is odd with respect to a $\mathbb{Z}_2$-grading and by

$$\mathrm{ind}_G(D) := \partial_1\left[\frac{\chi(D)+1}{2}\right] \in K_1(C^*(M)^G) \cong K_1(C_r^*(G))$$

when D is ungraded. Here,

$$\partial_i : K_{i+1}(D^*(M)^G/C^*(M)^G) \to K_i(C^*(M)^G) \cong K_i(C_r^*(G))$$

is the connecting maps of the K-theory six-term exact sequence associated with the short exact sequence:

$$0 \to C^*(M)^G \to D^*(M)^G \to D^*(M)^G/C^*(M)^G \to 0.$$

Let $K_i^G(M)$ be the equivariant K-homology of M. It captures the feature of canonical Dirac operators on M of dimension i (mod 2). For every Dirac-type operator D,

$$[\chi(D)] \in K_*^G(M).$$

The K-homology class does not depend on the choice of the normalizing function χ. See [30].

In general, there exits a functorial map

$$K_i^G(M) \to K_i(C_r^*(G))$$

sending $[\chi(D)]$ to the higher index $\mathrm{ind}_G(D) = \partial_i[\chi(D)]$. This higher index map is crucial in the formulation of the Baum–Connes assembly map. We will review the definition of higher indices using KK-theory in Section 3.4.13.

Example 3.4.12. Consider the manifold $\mathbb{R}^2$ acted by the additive group $\mathbb{R}^2$ by translation. Let D be the Dolbeault operator on $\mathbb{R}^2$. It is not a Fredholm operator. However, regard this as an $\mathbb{R}^2$-invariant operator, it has a higher index in $K_0(C_r^*(\mathbb{R}^2))$. Noting that $C_r^*(\mathbb{R}^2)$ is isomorphic to $C_0(\mathbb{R}^2)$ by the Gelfand transform, we shall find the higher index in $K_0(C_0(\mathbb{R}^2))$.

First, under the Gelfand transform, the Dolbeault operator D on $\mathbb{R}^2$ can be identified as $D = \begin{bmatrix} 0 & \bar{z} \\ z & 0 \end{bmatrix}$ as a family of operators over $z \in \mathbb{R}^2$. Choose $\chi(x) = \frac{x}{\sqrt{1+|x|^2}}$, and then

$$\chi(D) = \begin{bmatrix} 0 & \frac{\bar{z}}{1+|z|^2} \\ \frac{z}{1+|z|^2} & 0 \end{bmatrix}.$$

This represents an element in $K_1(C_b(\mathbb{R}^2)/C_0(\mathbb{R}^2))$. Lift $\chi(D)$ to an invertible element in $M_2(C_b(\mathbb{R}^2))$:

$$L = \begin{bmatrix} \frac{1}{|1+|z|^2|} & \frac{\bar{z}}{1+|z|^2} \\ \frac{z}{1+|z|^2} & \frac{1}{1+|z|^2} \end{bmatrix}$$

and then by definition of the K-theory boundary map, we have

$$\partial[\chi(D)] = L\begin{bmatrix} 1 & 0 \\ 0 & 0 \end{bmatrix}L^{-1} - \begin{bmatrix} 0 & 0 \\ 0 & 1 \end{bmatrix} = \frac{1}{1+|z|^2}\begin{bmatrix} 1 & \bar{z} \\ z & |z|^2 \end{bmatrix} - \begin{bmatrix} 0 & 0 \\ 0 & 1 \end{bmatrix} \in K_0(C_0(\mathbb{R}^2)).$$

Recall that

$$\beta := \frac{1}{1+|z|^2}\begin{bmatrix} 1 & \bar{z} \\ z & |z|^2 \end{bmatrix} - \begin{bmatrix} 0 & 0 \\ 0 & 1 \end{bmatrix} \in K_0(C_0(\mathbb{R}^2))$$

is the well-known Bott generator. Hence, we have shown that the higher index of D is a nontrivial element generating $K_0(C_r^*(\mathbb{R}^2))$. This can be regarded as the family index of D over the Pontryagin dual $\mathbb{R}^2$ of $\mathbb{R}^2$.

The "noncommutative" analogue of the Fourier transform performed in this example is crucial in finding generators for the K-theory of $C_r^*(G)$ for a connected reductive Lie group.

3.4.13 Higher Index via Hilbert C^*-Modules

Let M be a complete Riemannian manifold. Consider a Dirac-type operator D on $S \to M$ that is essentially self-adjoint. Let G be an almost connected Lie group acting on M properly, cocompactly, and ismorphically. Similar to Subsection 3.4.12, D can be regarded as a generalized Fredholm operator relative to a Hilbert module over $C_r^*(G)$. The construction is well known from [33]. Refer to Section 5 of [35] for full details.

First, let $C_c^\infty(G)$ be the convolution $*$-algebra of locally compact smooth functions on G, and regard $C_c^\infty(M)$ as a $C_c^\infty(G)$-module: For $\xi \in C_c^\infty(M)$ and $f \in C_c^\infty(G)$

$$(\xi * f)(x) = \int_G \xi(xg^{-1}) f(g) dg \qquad x \in M.$$

Moreover, $C_c^\infty(M)$ admits a $C_c^\infty(G)$-valued inner product $\langle \xi, \eta \rangle_G \in C_c^\infty(G)$ defined by

$$\langle \xi, \eta \rangle_G : g \mapsto \int_M \langle \xi(x), \eta(xg) \rangle dx$$

for $\xi, \eta \in C_c^\infty(M)$. The inner product is compatible with the module structure in the following sense. For $\xi, \eta \in C_c^\infty(M)$ and $f \in C_c^\infty(G)$, one has

$$\langle \xi, \eta * f \rangle_G = \langle \xi, \eta \rangle_G * f.$$

Definition 3.4.14. The completion of $C_c^\infty(M)$ under the inner product $\langle \cdot, \cdot \rangle_G$ is a right Hilbert module over $C_r^*(G)$, denoted by $\mathscr{E}_G(M)$. The completion can be applied similarly to $C_c^\infty(M,S)$ to obtain a $C_r^*(G)$-module $\mathscr{E}_G(M,S)$.

Let A be a C^*-algebra and $\mathscr{E}$ a Hilbert A-module. Denote by $\mathscr{L}(\mathscr{E})$ the C^*-algebra of bounded adjointable operators on $\mathscr{E}$ and $\mathscr{K}(\mathscr{E})$ the subalgebra of compact operators. Note that the compact operator $\mathscr{K}(\mathscr{E})$ is Morita equivalent to A, hence it induces an isomorphism of K-theory:

$$K_*(\mathscr{K}(\mathscr{E})) \cong K_*(A).$$

Remark 3.4.15. *By the Kasparov stabilization theorem, a separable Hilbert A-module $\mathscr{E}$ is a direct summand of the free Hilbert A-module $H_A := A \otimes H$ and the compact operators $\mathscr{K}(\mathscr{E})$ is a subalgebra of $\mathscr{K}(H_A)$, which can be identified as $A \otimes \mathscr{K}(H)$.*

Let χ be a normalizing function as in (3.4.1.1), so that $\hat{\chi}$ has compact support. Then the bounded operator $\chi(D)$ has finite propagation. In this setting, $\chi(D)$ is a pseudo-differential operator of order 0 with proper support. Thus, applying [35, Propositions 5.4–5.5], we have:

Proposition 3.4.16. *Let D be a Dirac-type operator on M. Then*

$$\chi(D) \in \mathscr{L}(\mathscr{E}_G(M,S)) \qquad 1 - \chi(D)^2 \in \mathscr{K}(\mathscr{E}_G(M,S)).$$

Definition 3.4.17. The *higher index* of D (odd $\mathbb{Z}_2$-graded) is defined by

$$\mathrm{ind}_G(D) := \partial_0[\chi(D)] \in K_0(\mathcal{K}(\mathcal{E}_G(M,S))) \cong K_0(C_r^*(G)).$$

Here $\partial_0 : K_1(\mathcal{L}(\mathcal{E}_G(M,S))/\mathcal{K}(\mathcal{E}_G(M,S))) \to K_0(\mathcal{K}(\mathcal{E}_G(M,S)))$ is the boundary map of the K-theoretic six-term exact sequence associated with the short exact sequence

$$0 \to \mathcal{K}(\mathcal{E}_G(M,S)) \to \mathcal{L}(\mathcal{E}_G(M,S)) \to \mathcal{L}(\mathcal{E}_G(M,S))/\mathcal{K}(\mathcal{E}_G(M,S)) \to 0.$$

The last isomorphism follows from the fact that $\mathcal{K}(\mathcal{E}_G)$ is Morita equivalent to $C_r^*(G)$. In the ungraded case or in the case of a canonical Dirac operator on M with odd dimension, one can similarly define the higher index

$$\mathrm{ind}_G(D) := \partial_1\left[\frac{\chi(D)+1}{2}\right] \in K_1(\mathcal{E}_G(M,S)) \cong K_1(C_r^*(G)).$$

Remark 3.4.18. *Suppose D is odd with respect to a $\mathbb{Z}_2$-grading. Because D is G-invariant, then $g \ker D^\pm \subset \ker D^\pm$ for all $g \in G$. So $\ker D^\pm$ is a representation space of G, and then by the universal property of the maximal group C^*-algebra, $\ker D^+$ and $\ker D^-$ are also $C^*(G)$-modules. Up to a compact perturbation, the higher index is the formal difference of two modules:*

$$\mathrm{ind}_G(D) = [\ker(\chi(D)^+ + K_0)] - [\ker(\chi(D)^- + K_1)]$$

where K_i are compact operators on the Hilbert module $\mathcal{E}_G(M)$. See Chapter 17 of [52]. In particular, if G is a compact Lie group acting on a closed manifold M and D is a G-invariant odd Dirac-type operator, then the higher index of D coincides with the equivariant index

$$\mathrm{ind}_G(D) := [\ker D^+] - [\ker D^-] \in K_0(C_r^*(G)).$$

This shows that the higher index is a generalization of the equivariant index when G is compact.

To understand the higher index from the point of view of noncommutative harmonic analysis, it is important to simplify the expression of the higher index in the context of KK-theory. In fact, it would be more elegant to use the KK-theory picture to describe the higher index map. Note that the isomorphism

$$KK_i(\mathbb{C}, C_r^*(G)) \cong K_i(C_r^*(G)) \tag{3.4.18.1}$$

is given exactly by the boundary maps

$$[(\mathcal{E}, F)] \mapsto \partial_0[F] \text{ or } \partial_1[(F+1)/2] \in K_i(\mathcal{K}(\mathcal{E})).$$

Therefore, we can rewrite

$$\mathrm{ind}_G(D) = [(\mathcal{E}_G, \chi(D))] \in KK_i(\mathbb{C}, C_r^*(G)).$$

Remark 3.4.19. *In the even ($\mathbb{Z}_2$-graded) case of i, $[(\mathcal{E}_G(M,S),\chi(D))]$ can be written as*

$$[(\mathcal{E}_G(M,S^+),\mathcal{E}_G(M,S^-),(1+D^+D^-)^{-\frac{1}{2}}D^+)] \in KK_0(\mathbb{C},C_r^*(G)).$$

Note that here we chose $\chi(x) = \frac{x}{\sqrt{x^2+1}}$ and one can show that D is a regular essential self-adjoint operator on the $C_r^*(G)$-module $\mathcal{E}_G$ with $\chi(D)^2 - I \in \mathcal{K}(\mathcal{E}_G)$. Thus, the KK cycle is well defined. This is an example of the alternative picture of $K_0(A)$ by operators on A-modules.

Note in addition that $\mathcal{E}_G$ is the direct summand of $L^2(G,L^2(M,E))$ obtained by convolving the canonical projection $p \in C_0(M) \rtimes G$ defined by

$$[p(g)](x) := c(x)(g\cdot c)(x) \tag{3.4.19.1}$$

where $c \in C_c(M)$ is a cutoff function satisfying $\int_G (g\cdot c)^2(x)dg = 1$ for all $x \in M$. See Section 5 of [35]. Therefore, one has the following equivalent definition of the higher index map:

Definition 3.4.20 ([33])**.** Let G be a locally compact group acting properly, cocompactly, and isometrically on a complete Riemannian manifold M. The *higher index map*

$$\mathrm{ind}_G : K_i^G(M) \to K_i(C_r^*(G))$$

is given by

$$\mathrm{ind}_G[F] := [p] \otimes_{C_0(M)\rtimes G} j^G[F]$$

where $p \in C_0(M) \rtimes G$ is the canonical projection in (3.4.19.1) and

$$j^G : KK_i^G(A,B) \to KK_i(A\rtimes G, B\rtimes G)$$

is the descent map. In particular, let D be a Dirac-type operator on M, then

$$\mathrm{ind}_G[\chi(D)] = [(\mathcal{E}_G(M),\chi(D))] \in KK_i(\mathbb{C},C_r^*(G)). \tag{3.4.20.1}$$

Example 3.4.21. Let G be an almost connected Lie group and $M = G/K$ the homogeneous space where K is a maximal compact subgroup of G. Suppose that M has a G-equivariant spin structure and Δ is the irreducible representation of $\mathrm{Spin}(\mathfrak{p})$. Let V be an irreducible representation of K. Formulate the balanced product

$$E = G\times_K(\Delta\otimes V)$$

where K acts on Δ through $K \to \mathrm{Spin}(\mathfrak{p}) \to \mathrm{End}(\Delta)$. Then one has $\mathcal{E}_G(M,E) = (L^2(G)\otimes\Delta\otimes V)^K$ and

$$\mathcal{K}(\mathcal{E}_G) \subset C_r^*(G)\otimes\mathrm{End}(\Delta\otimes V).$$

If G is abelian or $C_r^*(G)$ has the nice property of being Morita equivalent to $C_0(\widehat{G}_t)$ (for example, when G is complex semisimple), then

$$\mathrm{ind}_G(D) \in K_n(C_r^*(G)) \cong K_n(C_0(\widehat{G}_t)).$$

We shall see that this higher index is the index of a family of operators $\{D_\pi\}_{\pi\in\widehat{G}_t}$ where D_π is D restricted to $\mathcal{E}_G(X,E) \otimes_{C_r^*(G)} H_\pi$, the component associated to $(\pi, H_\pi) \in \widehat{G}_t$. Compare Example 3.4.12.

The idea of using family indices in Examples 3.4.12 and 3.4.21 is crucial in calculating $K_*(C_r^*(G))$ and proving the Connes–Kasparov isomorphism when G is connected, semisimple, or reductive. When decomposing an operator on a Hilbert $C_r^*(G)$-module into a family of operators, one needs to apply some noncommutative Fourier transform to the operator. Thus, in that case, it would be more convenient to use the unbounded picture instead of bounded cycles in (3.4.20.1). It would then be natural to ask for an elliptic differential operator to represent a KK-cycle. This is not a problem for an almost connected Lie group G. In fact, it is well known that all G-equivariant K-homology classes of a proper cocompact manifold M can be represented by twisted Dirac operators D, in the sense of Baum–Douglas geometric K-homology [10]. Also, the KK-cycle $(\mathcal{E}_G, \chi(D))$ representing the higher index can be replaced by the unbounded KK-cycle $(\mathcal{E}_G, D)$ in the sense of Baaj-Julg [11]. See also [37, 42]. One can also see it through the fact that under the isomorphism (3.4.18.1), higher indices of D and $\chi(D)$ coincide in $K_*(C_r^*(G))$, because the higher index under the boundary map is independent of the choice of a parametrix. The higher index map in Definition 3.4.20 can then be written as

$$\mathrm{ind}_G : K_i^G(M) \to KK_i(\mathbb{C}, C_r^*(G)) \qquad [D] \mapsto [(\mathcal{E}_G, D)] \tag{3.4.21.1}$$

where D is a twisted Dirac operator on a manifold carrying a proper and cocompact G action. When the Clifford module where D acts on has a $\mathbb{Z}_2$-grading and when D is odd, $i = 0$. For the ungraded case, $i = 1$.

Example 3.4.22. Let D be the Dolbeault operator on $\mathbb{R}^2$, as an $\mathbb{R}^2$-invariant operator. We have shown its higher index is given by the Bott element: $\beta = \frac{1}{1+|z|^2}\begin{bmatrix} 1 & \bar{z} \\ z & |z|^2 \end{bmatrix} - \begin{bmatrix} 0 & 0 \\ 0 & 1 \end{bmatrix} \in K_0(C_0(\mathbb{R}^2))$. Corresponding element in $KK(\mathbb{C}, C_0(\mathbb{R}^2))$ is represented by

$$\left(C_0(\mathbb{R}^2, \mathbb{C}^2), \begin{bmatrix} 0 & \bar{z} \\ z & 0 \end{bmatrix}\right)$$

in the unbounded KK-model or

$$\left(C_0(\mathbb{R}^2), C_0(\mathbb{R}^2), \frac{z}{1+|z|^2}\right)$$

in the ordinary KK-model. Thus, under the higher index map

$$\mathrm{ind}_G : K_0^{\mathbb{R}^2}(\mathbb{R}^2) \to KK_0(\mathbb{C}, C_r^*(\mathbb{R}^2)) \cong KK_0(\mathbb{C}, C_0(\mathbb{R}^2)),$$

the unbounded picture of the higher index is given by

$$\mathrm{ind}_G[D] = \left[\left(C_0(\mathbb{R}^2, \mathbb{C}^2), \begin{bmatrix} 0 & \bar{z} \\ z & 0 \end{bmatrix}\right)\right] \in KK_0(\mathbb{C}, C_0(\mathbb{R}^2)).$$

Example 3.4.23. Note that

$$\begin{aligned} \begin{bmatrix} 0 & \bar{z} \\ z & 0 \end{bmatrix} = \begin{bmatrix} 0 & x_1 - ix_2 \\ x_1 + ix_2 & 0 \end{bmatrix} &= ix_1 \begin{bmatrix} 0 & -i \\ -i & 0 \end{bmatrix} + ix_2 \begin{bmatrix} 0 & -1 \\ 1 & 0 \end{bmatrix} \\ = \quad & ix_1 c(e_1) + ix_2 c(e_2) = ic(x). \end{aligned}$$

One can generalize the above description to $\mathbb{R}^n$. The generator for $KK_n(\mathbb{C}, C_0(\mathbb{R}^n))$ is given by

$$\left(C_0(\mathbb{R}^n, \mathbb{C}^{2^{\lfloor n/2 \rfloor}}), ic(x)\right) \in KK_n(\mathbb{C}, C_0(\mathbb{R}^n)).$$

Here, $x = x_1e_1 + x_2e_2 + \cdots + x_ne_n \in \mathbb{R}^n$. This presentation of the Bott element is often used in the context of E-theory.

The higher index map for the canonical Dirac operator D on $\mathbb{R}^n$ under the higher index map

$$\mathrm{ind}_G : K_n^{\mathbb{R}^n}(\mathbb{R}^n) \to KK_n(\mathbb{C}, C_r^*(\mathbb{R}^n)) \cong KK_n(\mathbb{C}, C_0(\mathbb{R}^n))$$

is given by

$$\mathrm{ind}_G[D] = \left[\left(C_0(\mathbb{R}^n, \mathbb{C}^{2^{\lfloor n/2 \rfloor}}), ic(x)\right)\right].$$

3.5 Connes–Kasparov Isomorphism

For a locally compact group G, calculating the K-theory of its reduced group C^*-algebra is an important but difficult problem. Baum and Connes formulated an algorithm of computing the K-theory, known as the Baum–Connes conjecture [8, 9]. When G is an almost connected Lie group, the conjecture is known as the Connes–Kasparov conjecture, and has been verified to be true. The conjecture states that the Baum–Connes assembly map is an isomorphism for G. The Connes–Kasparov isomorphism restricted to more classical groups, such as semisimple or reductive Lie groups, has generated amazingly fruitful communications between representation theory and operator algebras. This is because the isomorphism is intimately related to the representation theory of a class of noncompact Lie groups and their geometric construction via Dirac operators.

In this section, we will recall relevant ideas involving the Connes–Kasparov conjecture, emphasizing motivations coming from representation theory and Dirac operators. We will first introduce Dirac induction, formulating the Connes–Kasparov morphism, followed by some examples. Then we briefly introduce the Baum–Connes assembly map, including the Connes–Kasparov morphism as a special case, and the deformation-to-the-normal-cone construction of the assembly map, which is closely related to the K-theoretic Mackey bijection initiated by Higson [29]. Finally, we go over the idea of the Connes–Kasparov isomorphism via the Dirac-dual Dirac method, which is one of the classical ways of proving the Baum–Connes conjecture [34, 39]. We shall review the representation-theoretic proof in the last section, via the noncommutative Fourier transform and harmonic analysis on nice Lie groups.

3.5.1 Dirac Induction

Following the seminal work [26, 27] by Harish-Chandra regarding discrete series representations of a real semisimple Lie group, various geometric constrictions of representations of a connected noncompact semisimple Lie group G have been developed, generalizing the Borel–Weil–Bott theory for compact Lie groups. Considering Dolbeault operators on a noncompact homogeneous space of the form G/T, where T is a compact Cartan subgroup of G, Schmid [46] constructed most discrete series using the Dolbeault L^2-coholomogy of G. Parthasarathy [43] then used the space of L^2-solutions of general Dirac-type operators on the homogeneous space G/K, where K is a maximal compact subgroup of G, to produce discrete series for G. Atiyah–Schmid [5] pushed the idea further by explicitly working out the parametrization of each Dirac-type operator whose L^2-kernel represents a discrete series representation. Noting the higher index of a G-invaraint Dirac-type operator D is essentially given by the compact perturbation of the projection onto its L^2-kernel. The work by Atiyah–Schmid strongly motivates the explicit description of the Connes–Kasparov morphism, through the so-called Dirac induction, which we will now introduce.

Let us assume now that G is an almost connected Lie group and K is a maximal compact subgroup. Consider the noncompact homogeneous space G/K, equipped with a G-equivariant spin structure. Let $S \to G/K$ be the spinor bundle and

$$D : C_c^\infty(G/K, S) \to C_c^\infty(G/K, S)$$

be the Dirac operator associated to the spin structure. In addition, associated with a representation (ρ, V) of K, one can construct the homogeneous vector bundle

$$\mathcal{V} := G \times_K V \tag{3.5.1.1}$$

over G/K obtained by taking the quotient of $G \times V$ by the K-action:

$$k \cdot (g, v) = (gk, \rho(k^{-1})v) \qquad k \in K, v \in V, g \in G.$$

Then one may form the twisted Dirac operator

$$D_{\mathcal{V}} : C_c^\infty(G/K, S \otimes \mathcal{V}) \to C_c^\infty(G/K, S \otimes \mathcal{V}).$$

Let $R(K)$ be the representation ring generated by irreducible unitary representations of K.

Definition 3.5.2. The *Dirac induction* is the morphism of abelian groups

$$R(K) \to K_i(C_r^*(G)) \tag{3.5.2.1}$$

sending $[V] \in R(K)$ to the higher index $\mathrm{ind}_G(D_{\mathcal{V}}) \in K_i(C_r^*(G))$. Here, i is the parity of $\dim(G/K)$.

Remark 3.5.3. *If G/K does not have a G-equivariant spin structure, then construct a double cover $\widetilde{K}$ of K as in [22]: $\widetilde{K}$ is the subgroup of $K \times \mathrm{Spin}(\mathfrak{p})$ so that $(h, s) \in \tilde{K}$ if and only if $Ad^*(h)$ is equal to the orthogonal transformation of $\mathfrak{p}$ defined by $s \in \mathrm{Spin}(\mathfrak{p})$. Then the map $\widetilde{K} \to SO(\mathfrak{p})$ given by $(h, s) \mapsto Ad^*(h)$ lifts to $\widetilde{K} \to \mathrm{Spin}(\mathfrak{p})$ naturally. One can then still formulate the Dirac induction, replacing $R(K)$ by*

$$R_{\mathrm{spin}}(K) := \{\pi \in R(\widetilde{K}) \text{ that does not factor through } K\}.$$

The subring of $R(\widetilde{K})$ consisting of elements that would factor through K can be identified with $R(K)$. In fact, $\widehat{\widehat{K}}$ contains $\widehat{K}$ and $R(\widetilde{K}) = R(K) \oplus R_{\mathrm{spin}}(K)$.

Theorem 3.5.4 (Connes–Kasparov isomorphism)**.** *Let G be an almost connected Lie group and K a maximal compact subgroup. The Dirac induction map*

$$R(K) \to K_*(C_r^*(G)) \qquad [V] \mapsto \mathrm{ind}_G(D_V)$$

is an isomorphism of abelian groups. When G is connected, the total K group on the right-hand side can be replaced by $K_n(C_r^*(G))$ where $n = \dim(G/K)$.

Let us recall some elements of Harish-Chandra and Atiyah–Schmid's work in order to see that the Connes–Kasparov isomorphism is in fact related to Atiyah–Schmid's geometric way of construction of discrete series representations. Let G be a connected semisimple Lie group with discrete series. Let π_λ be the discrete series labeled by the Harish-Chandra parameter $\lambda \in i\mathfrak{t}^*$, that is,

- $(\lambda, \alpha) \neq 0$ for all $\alpha \in R(\mathfrak{t}, \mathfrak{g})$;
- $\lambda + \rho$ is analytic integral.

Here, $R(\mathfrak{t}, \mathfrak{g})$ is the set of roots for $(\mathfrak{t}, \mathfrak{g})$, and ρ is the half sum of positive roots. Recall that a discrete series representation is contained in the regular representation $L^2(G)$ as a direct summand. Thus the associated projection onto the direct summand represents an element of $K_0(C_r^*(G))$. Then the main result by Atiyah–Schmid is formulated in the language of K-theory as follows:

Theorem 3.5.5 ([5, Theorem 5.20], see also [9, Theorem 4.26]). *The Dirac induction*

$$R(K) \to K_0(C_r^*(G))$$

sends $E = E_\tau \in \widehat{K}$ with the highest weight τ to the discrete series generator with Harish-Chandra parameter $\tau + \rho_K$ where ρ_K is the half sum of compact positive roots.

Atiyah–Schmid's result verifies the Connes–Kasparov isomorphism when the group G is connected semisimple having discrete series (when $\mathrm{rank}(G) = \mathrm{rank}(K)$) and over the discrete series generators, which does *not* constitute all K-theory generators.

Example 3.5.6. Consider $G = SL(2, \mathbb{R})$, and then $K = SO(2)$. Note that $\mathrm{rank}(G) = \mathrm{rank}(K)$, so G has discrete series. Label the discrete series by $D_1^\pm, D_2^\pm, \ldots$. Because $\dim G/K = 2$, we shall compare:

$$R(SO(2)) = \bigoplus_{n \in \mathbb{Z}} \mathbb{Z};$$

$$K_0(C_r^*(SL(2,\mathbb{R}))) \cong \left(\bigoplus_{n \neq 0} \mathbb{Z}\right) \bigoplus K_0(C_0(\mathbb{R}) \rtimes \mathbb{Z}_2) \oplus K_0(C_0(\mathbb{R}/\mathbb{Z}_2)).$$

The contribution of the generator associated with the label n in $R(SO(2))$ is (μ, V_μ) with the highest weight $\mu = n$. The contribution of the generator associated with the label $n \neq 0$ in $K_0(C_r^*(SL(2,\mathbb{R})))$ is represented by the dis-

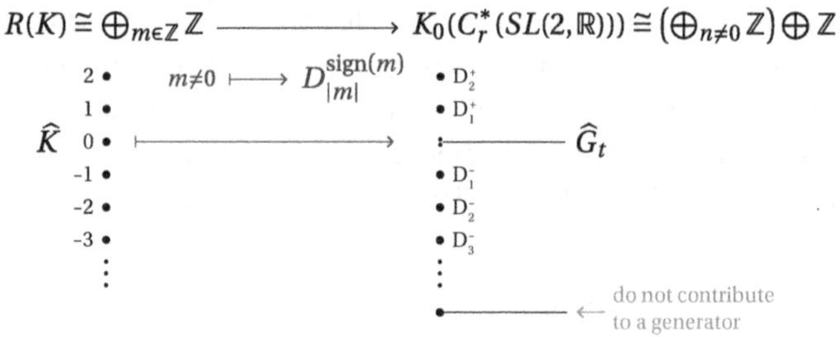

$R(K) \cong \bigoplus_{m\in\mathbb{Z}} \mathbb{Z} \longrightarrow K_0(C_r^*(SL(2,\mathbb{R}))) \cong (\bigoplus_{n\neq 0}\mathbb{Z}) \oplus \mathbb{Z}$

$m\neq 0 \longmapsto D_{|m|}^{\text{sign}(m)}$

$\widehat{K}$: 2, 1, 0, -1, -2, -3, … $\longmapsto$ D_2^+, D_1^+, D_1^-, D_2^-, D_3^-, … — $\widehat{G}_t$

do not contribute to a generator

Figure 3.1 Connes–Kasparov correspondence for $SL(2, \mathbb{R})$

crete series $D_{|n|}^{\text{sign}n}$ with Harish-Chandra parameter $\frac{n\alpha}{2}$ where α is the positive root of the Cartan subgroup $T = K = SO(2)$. Note that $\rho_K = 0$, then, by Atiyah–Schmid,

$$D_{|n|}^{\text{sign}n} = \ker D_{V_n}^+ = \text{ind}_G(D_{V_n}^+).$$

See Figure 3.1. Note that the parametrization is slightly different from Example 4.25 of [9]. See an explanation for the different parametrizations in Remark 6 of [28].

Besides discrete series, components of irreducible principal series in the tempered dual of a connected semisimple Lie group is also an important source for the K-theory generators. Penington–Plymen studied the case of a complex semisimple Lie group, which has no discrete series and whose K-theory generators are all made up of connected components of irreducible principal series. Recall that the tempered dual $\widehat{G}_t$ has no discrete series and its Fell topology is Hausdorff. Being more precise, via parabolic induction, $\text{Ind}_{P=MAN}^G(\sigma\otimes\lambda\otimes 1)$ where $\sigma \in \widehat{M}, \lambda \in \widehat{A}$, one has the identification:

$$\widehat{G_t} = (\widehat{M}\times\widehat{A})/W.$$

where P is the minimal parabolic subgroup of G, $P = MAN$ is the Langlands decomposition, and W is the Weyl group. For $\sigma \in \widehat{M}$, the stabilizer

$$W_\sigma := \{w \in W | w\sigma = \sigma\}$$

acts on $\{\sigma\}\times\widehat{A}$. Thus

$$\widehat{G}_t = (\widehat{M}\times\widehat{A})/W = \bigcup_{[\sigma]\in\widehat{M}/W} (\{\sigma\}\times(\widehat{A}/W_\sigma)).$$

Here, $\{\sigma\}\times(\widehat{A}/W_\sigma)$ is called the component labeled by σ, and contributes a nontrivial generator in K-theory if and only if W_σ is trivial:

$$K_i(C_r^*(G)) = K^i(G_t) = K^i((\widehat{M}\times\widehat{A})/W) = \bigoplus_{W_\sigma=1} K^i(\{\sigma\}\times\widehat{A}) \quad i = \dim A.$$

Theorem 3.5.7 ([44]). *Let G be a complex connected semisimple Lie group. The Dirac induction is an isomorphism*

$$R(K) \to K_i(C_r^*(G)) \qquad i = \dim G/K$$

sending $[E_\tau] \in \widehat{K}$ *the irreducible representation of K with highest weight* τ *to the component for* $\sigma \in \widehat{M}$ *with highest weight* $\tau + \rho_K$*, where* ρ_K *is the half sum of compact positive roots.*

Example 3.5.8. Consider $G = SL(n, \mathbb{C})$, with the maximal compact subgroup $K = SU(n)$. Because complex semisimple Lie groups do not have discrete series, while in the structure of $C_r^*(G)$, the component labeled by $[P, \sigma]$ requires that $P = MAN$ is parabolic and $\sigma \in \widehat{M}_d$, that is, σ is a discrete series of M, so we only need to consider the minimal parabolic subgroup $P = MAN$, the subgroup consisting of upper triangular matrices. Here

$$M = \left\{ \begin{bmatrix} e^{i\theta_1} & & \\ & \ddots & \\ & & e^{i\theta_{n-1}} \end{bmatrix} \mid \theta_k \in \mathbb{R}, \theta_1 + \cdots + \theta_{n-1} = 0 \right\}$$

and

$$A = \left\{ \begin{bmatrix} x_1 & & \\ & \ddots & \\ & & x_{n-1} \end{bmatrix} \mid x_k \in \mathbb{R}, x_1 \cdots x_{n-1} = 1 \right\}.$$

So $\widehat{M} = \mathbb{Z}^{n-1}$ and $\widehat{A} \cong \mathbb{R}^{n-1}$. The Weyl group W is the subgroup generated by elementary matrices of the form $R_{m,n}$, which represents exchanging of rows m and n, and the Weyl group acts on A, M by conjugations. Hence, W is isomorphic to S_n, the symmetric group on n numbers. Therefore,

$$K_{n-1}(C_r^*(SL(n, \mathbb{C}))) \cong K_{n-1}((\widehat{M} \times \widehat{A})/S_n) \cong \bigoplus_{0 < m_1 < \cdots < m_{n-1}} \mathbb{Z}.$$

Recall that for the compact Lie group K, the representation ring $R(K)$ is generated by $\widehat{K} = \widehat{T}^W$ where in this example, T is the maximal torus

$$T = \left\{ \begin{bmatrix} e^{i\theta_1} & & \\ & \ddots & \\ & & e^{i\theta_{n-1}} \end{bmatrix} \mid \theta_k \in \mathbb{R}, \theta_1 + \cdots + \theta_{n-1} = 0 \right\}$$

on which $W = N_K(T)/T \cong S_n$ permutes the diagonal elements. Thus,

$$R(SU(n)) \cong \bigoplus_{0 \le m_1 \le \cdots \le m_{n-1}} \mathbb{Z}$$

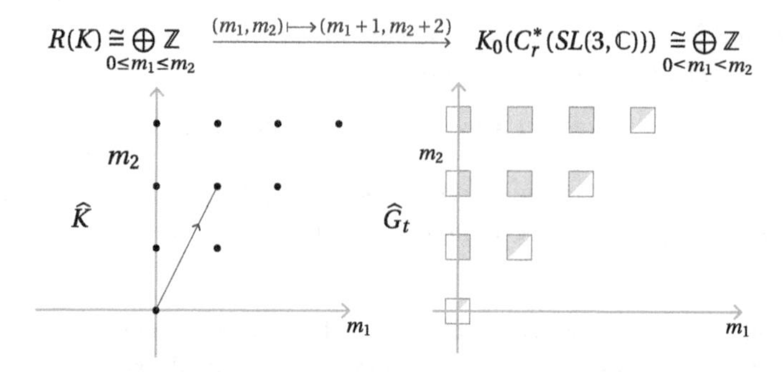

Figure 3.2 Connes–Kasparov isomorphism for $SL(3, \mathbb{C})$

By Theorem 3.5.7 due to Penington and Plymen, the generator $\tau \in \widehat{K}$ with the highest weight

$$(m_1, \ldots, m_{n-1}) \in \mathbb{Z}^{n-1} \cong \widehat{T}$$

where $0 \le m_1 \le \cdots \le m_{n-1}$, corresponds the K-theory component labeled by $\sigma \in \mathbb{Z}^{n-1} = \widehat{M}$ with the highest weight

$$\tau + \rho_K = (m_1 + 1, m_2 + 2, \ldots, m_{n-1} + n - 1).$$

See Figure 3.2 for the correspondence for $SL(3,\mathbb{C})$. Note that neither the closed half spaces nor the 1/6 space in the reduced dual of $SL(3,\mathbb{C})$ in the figure contribute nontrivial K-theory elements.

The result by Penington–Plymen was then carried forward by Valette in a slightly general setting. Let G be a semisimple Lie group having one conjugacy class of Cartan subgroups. This includes the complex semisimple case, but also the example of $SL(3,\mathbb{R})$. This does not include $SL(2,\mathbb{R})$ because $SL(2,\mathbb{R})$ has compact and noncompact Cartan subgroups that cannot be conjugate to each other. In this setting, $\widehat{G_t}$ remains Hausdorff and

$$\widehat{G_t} \cong \bigsqcup_{P=MAN} (\widehat{M} \times \widehat{A})/W.$$

The parabolic subgroups P here in the disjoint union need to satisfy the requirement that M has discrete series.

Theorem 3.5.9 ([49]). *Let G be a connected semisimple Lie group having one conjugacy class of Cartan subgroups. The Dirac induction is an isomorphism*

$$R(K) \to K_i(C_r^*(G)) \qquad i = \dim G/K$$

sending $[E_\tau] \in \widehat{K}$ with the highest weight τ to the component for $\sigma \in \widehat{M}$ with the highest weight $\tau + \rho_K - \rho_M$, where ρ_K is the half sum of compact positive roots and ρ_M is the half sum of positive root of M.

Example 3.5.10 ([49]). Consider $G = SL(3,\mathbb{R})$, then the maximal compact subgroup $K = SO(3)$. There are *three* parabolic subgroups. The one $P = G$ does not contribute to the tempered dual because G has no discrete series. The other two parabolic subgroups have the form:

$$P_1 = \left\{ \begin{bmatrix} * & * & * \\ & * & * \\ & & * \end{bmatrix} \right\} \quad P_2 = \left\{ \begin{bmatrix} * & * & * \\ * & * & * \\ & & * \end{bmatrix} \right\}.$$

Under the Langlands decomposition $P_i = M_i A_i N_i$, we have

$$M_1 = \left\{ \begin{bmatrix} \epsilon_1 & 0 & 0 \\ 0 & \epsilon_2 & 0 \\ 0 & 0 & \epsilon_3 \end{bmatrix} | \epsilon = \pm 1, \epsilon_1\epsilon_2\epsilon_3 = 1 \right\}$$

$$A_1 = \left\{ \begin{bmatrix} x_1 & 0 & 0 \\ 0 & x_2 & 0 \\ 0 & 0 & x_3 \end{bmatrix} | x_i > 0, x_1x_2x_3 = 1 \right\}$$

and

$$M_2 = \left\{ \begin{bmatrix} a & b & 0 \\ c & d & 0 \\ 0 & 0 & e \end{bmatrix} | e = \pm 1, (ad - bc)e = 1 \right\} \quad A_2 = \left\{ \begin{bmatrix} x & 0 & 0 \\ 0 & x & 0 \\ 0 & 0 & x^{-2} \end{bmatrix} | x > 0 \right\}.$$

The Weyl group $N_G(L)/L$ where $L = MA$ is $W_1 = S_3$ for P_1 and trivial for P_2. The group W_1 acts on M_1, A_1 by permuting the diagonal elements. Hence

$$(\widehat{M}_1 \times \widehat{A}_1)/W_1 = (\mathbb{Z}_2^2 \times \mathbb{R}^2)/S_3$$

is the contribution of P_1 to the tempered dual, and it has two components, one is a closed half space and the other is a closed convex cone. They do not contribute any nontrivial elements in K-theory. For P_2,

$$\widehat{M}_{2,d} \times \widehat{A} = \mathbb{Z}_{\geq 1} \times \mathbb{R}$$

is the corresponding contribution to the reduced dual, and each component gives rise to a K-theory generator. Note that $M_2 \cong SL(2,\mathbb{R})^{\pm}$, the double cover of $SL(2,\mathbb{R})$, its discrete series is labeled by $n \geq 1$ given by D_n^+ or D_n^- induced from $SL(2,\mathbb{R})$. Therefore,

$$K_1(C_r^*(SL(3,\mathbb{R}))) \cong \bigoplus_{n \geq 1} \mathbb{Z}.$$

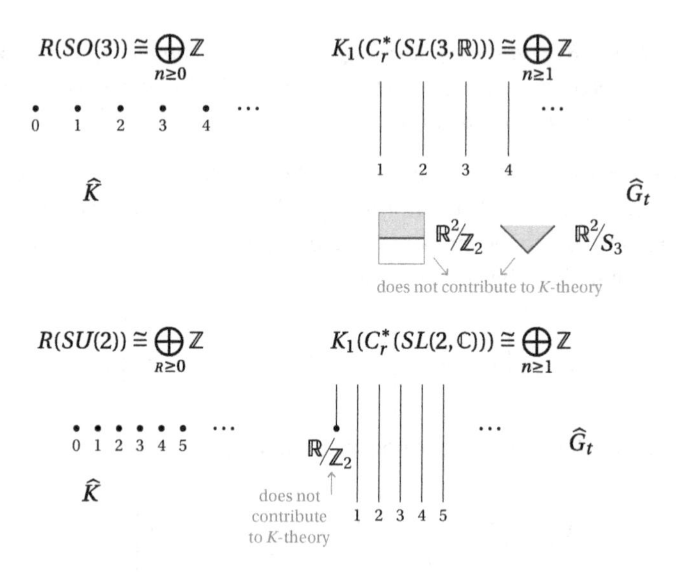

Figure 3.3 Connes–Kasparov isomorphism for $SL(3, \mathbb{R})$

To see $R(K)$, the irreducible representation of $SO(3)$ is labeled by $n \geq 0$ and

$$R(SO(3)) \cong \bigoplus_{n\geq 0} \mathbb{Z}.$$

According to [49], the Connes–Kasparov isomorphism is given by sending the component associated with n in $R(K)$ to the component associated with $n + 1$ in $K_1(C_r^*(SL(3,\mathbb{R})))$. The shift 1 of the labels is accommodated by $\rho_K - \rho_M$. Note that $SU(2)$ is a double cover of $SO(3)$, hence the amount of irreducible representations of $SU(2)$ is twice of those for $SO(3)$, so we also compare this example with the one for $SL(2,\mathbb{C})$. See Figure 3.3.

In 1987, Wassermann [51] applied the method of Fourier analysis over the reduced dual again and generalized the Connes–Kasparov isomorphism to connected linear reductive Lie groups. The paper is more technical but quite condensed. It contains for example real reductive groups where the topology of the reduced dual could involve non-Hausdorff point as a result of R-groups ($SL(2,\mathbb{R})$, for example). Recently, in [17], Clare–Crisp–Higson studied in greater detail the structure of the C^*-algebra of real reductive groups. Based on that, Clare, Higson, Tang, and Song reproved the Connes–Kasparov conjecture in [51] making use of Vogan's powerful method of minimal K-types [18, 19].

Theorem 3.5.11 ([18, 19, 51]). *Let G be a connected linear reductive Lie group. Then the Dirac induction is an isomorphism*

$$R(K) \to K_i(C_r^*(G)) \qquad i = \dim G/K.$$

Representation theory of an almost connect Lie group with more than one connected component is also very interesting. We discuss an example here. See also [41].

Example 3.5.12. Consider the group $G = GL(2,\mathbb{R})$, a real reductive group of two components. Its maximal compact subgroup is $K = O(2)$. Because $\mathrm{rank}(G) = \mathrm{rank}(K)$, the group has discrete series. The parabolic subgroups P_1, the subgroup of upper triangular matrices, and $P_2 = G$ contribute to the tempered dual. The corresponding M_i and A_i and Weyl groups are:

$$M_1 = \left\{ \begin{bmatrix} \epsilon_1 & 0 \\ 0 & \epsilon_2 \end{bmatrix} | \epsilon_i = \pm 1 \right\} \cong \mathbb{Z}_2^2 \quad A_1 = \left\{ \begin{bmatrix} x_1 & 0 \\ 0 & x_2 \end{bmatrix} | x_i > 0 \right\} \cong \mathbb{R}^2 \quad W_1 \cong \mathbb{Z}_2$$

and

$$M_2 = SL(2,\mathbb{R})^{\pm} \qquad A_2 = \left\{ \begin{bmatrix} x & 0 \\ 0 & x \end{bmatrix} | x > 0 \right\} \cong \mathbb{R} \qquad W_2 \cong \{e\}.$$

Thus, the tempered dual has the form

$$\widehat{G}_t = (\mathbb{Z}_2^2 \times \mathbb{R}^2)/\mathbb{Z}_2 \bigsqcup (\mathbb{Z}_{\geq 1} \times \mathbb{R}) = \mathbb{R}^2 \bigsqcup (\mathbb{R}^2/\mathbb{Z}_2) \bigsqcup (\mathbb{R}^2/\mathbb{Z}_2) \bigsqcup (\mathbb{Z}_{\geq 1} \times \mathbb{R})$$

where the first few components are principal series and the last component represents discrete series. Note that in the first three components, when $n_1 = n_2$ in $\mathbb{Z}_2^2$, we have the closed half plane $\lambda_2 \geq \lambda_1$; when $n_1 \neq n_2$, we have a full plane. The K-theory is then

$$K_1(C_r^*(GL(2,\mathbb{R}))) \cong \bigoplus_{n \geq 1} \mathbb{Z};$$
$$K_0(C_r^*(GL(2,\mathbb{R}))) \cong K_0(C_0(\mathbb{R}^2)) \cong \mathbb{Z}.$$

For $R(O(2))$, recall that irreducible representations consist of two *one*-dimensional representations: trivial one 1 and the sign representation sgn, and *two*-dimensional representations $\rho_n = \mathrm{Ind}_{SO(2)}^{O(2)} e_n$ for $n \geq 1$. Here, e_n is the irreducible representation of $SO(2)$ with the highest weight n. See Figure 3.4 for an illustration of $\widehat{K}$ and $\widehat{G}_t$. The precise match under the Connes–Kasparov isomorphism is left as an exercise. Note that by [23], the left-hand side of the Connes–Kasparov isomorphism

$$K_i^{O(2)}(V_2) \cong K_i(C_r^*(GL(2,\mathbb{R}))) \qquad i = 0, 1$$

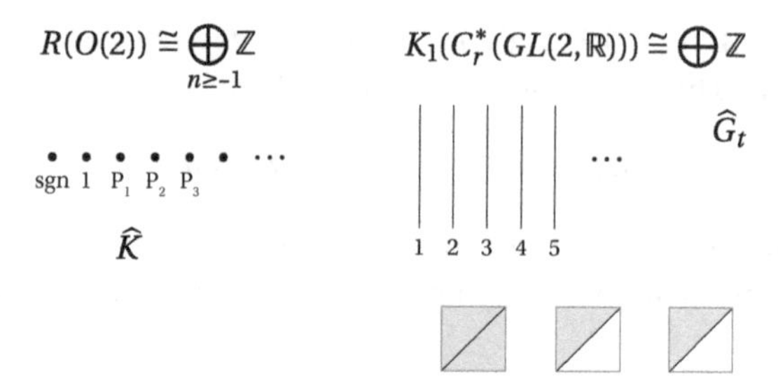

Figure 3.4 Connes–Kasparov isomorphism for $GL(2, \mathbb{R})$

is explicitly calculated. Here V_2 stands for the three-dimensional space of symmetric real 3×3 matrices, admitting the action of $O(2)$ by conjugation. One needs only to match generators of $R(O(2))$ with those of $K_0^{O(2)}(V_2) \oplus K_1^{O(2)}(V_2)$.

From a completely different angle, the Connes–Kasparov conjecture is proved for all almost connected Lie groups and linear p-adic groups. For the case of an almost connected Lie group, the injectivity was proved by the Dirac-dual Dirac method invented by Kasparov [34], where the feature that G/K being hyperbolic and negatively curved is playing an important role. The surjectivity was proved by Lafforgue [39], observing that the group has Property RD, so that the Dirac-dual Dirac method adapted to the setting of Banach KK-theory can be linked to ordinary K-theory of the reduced C^*-algebra. In [16], the Connes–Kasparov conjecture was proved in its most general setting, including the cases of an almost connected Lie group and a linear p-adic group, and with a new and powerful method by inducing from the action of some stabilizers on smaller pieces of the group.

Theorem 3.5.13 ([16, 34, 39]). *Let G be an almost connected Lie group or a linear p-adic group. Then the Dirac induction is an isomorphism*

$$R(K) \to K_*(C_r^*(G)).$$

The Connes–Kasparov isomorphism restricted to the cases of semisimple or reductive groups produces beautiful parametrization of representations, relating to the deep theory of harmonic analysis and representations of reductive groups, thanks to the pioneering work done by Harish-Chandra, Vogan, and many others. Moreover, the proofs of the cases concerning semisimple or reductive groups provide more explicit formulas for the K-theory of reduced group C^*-algebras. Therefore, the Connes–Kasparov isomorphism in the semisimple or reductive cases is valuable toward the connections between representation theory and noncommutative geometry.

3.5.14 Special Case of the Baum–Connes Conjecture

In this subsection, we briefly overview the Baum–Connes conjecture, where the Connes–Kasparov isomorphism is a special case.

Let us first recall the statement of the Baum–Connes conjecture. Let X be a topological space and G be a locally compact group. Recall that G acts on X properly if the preimage of every compact set under the following map is compact:

$$G \times X \to X \times X \qquad (g,x) \mapsto (g \cdot x, x).$$

Remark 3.5.15. *Locally, a proper G space is covered by G-slices of the form $G \times_H U$ where $H \subset G$ is a compact subgroup and $U \cong \mathbb{R}^k$ is an H-space.*

Let M be a topological space admitting a proper, cocompact action by a locally compact group G, the higher index map gives rise to a morphism

$$K_m^G(M) \to K_m(C_r^*(G)) \qquad [F] \to \mathrm{ind}_G(F) \tag{3.5.15.1}$$

where $m = 0$ or 1.

Definition 3.5.16. A contractible proper G-space X is a *universal example* if for every proper G-space M there is a unique G-equivariant map

$$M \to X$$

up to homotopy. Because X is unique up to homotopy, denote X by $\underline{E}G$, called the *universal space of proper actions* for G.

Remark 3.5.17. *The condition of being contractible can be removed because if M is universal, then M is contractible. See [9].*

Example 3.5.18. When G is an almost connected Lie group and K is a maximal compact subgroup, G/K is universal. In fact, the G-equivariant map from a proper G space to G/K locally has the form:

$$G \times_H U \to G/K \qquad [g,u] \mapsto [g].$$

Therefore, when G is an almost connected Lie group, G/K is a model for $\underline{E}G$. Clearly, G acts on G/K properly and cocompactly and G/K is contractible.

In general, a model for $\underline{E}G$ is a CW complex rather than a manifold and is not always G-cocompact.

For $m = 0$ or 1, denote by

$$K_m^G(\underline{E}G) = \lim_{X \to \underline{E}G} K_m^G(X)$$

where the limit is taken over all proper cocompact spaces X with respect to some G-equivariant maps $f : X \to \underline{E}G$.

Definition 3.5.19. The *Baum–Connes assembly map*

$$\mu : K_m^G(\underline{E}G) \to K_m(C_r^*(G))$$

is given by $f_*[F] \to \mathrm{ind}_G(F)$ where $[F] \in K_m^G(X)$ is represented by a G-invariant abstract elliptic operator F.

Conjecture 3.5.20 (Baum–Connes [8])**.** *The Baum–Connes assembly map μ is an isomorphism.*

Next, we show that the Dirac induction being an isomorphism is equivalent to the Baum–Connes conjecture for connected Lie groups.

Let G be a connected Lie group and K a maximal compact subgroup. The homogeneous space $M = G/K$ is a Riemannian manifold with a G-invariant metric given by the Killing form. Denote by $K_G^0(M)$ the Grothendieck group generated by the isomorphism class of G-equivariant vector bundles over M. In fact, they are homogeneous vector bundles of the form (3.5.1.1). The induction from a K-representation to a homogeneous vector bundle as a G-representation gives rise to an isomorphism

$$R(K) \to K_G^0(G/K) \qquad [V] \mapsto [\mathcal{V}] := [G \times_K V]. \tag{3.5.20.1}$$

Suppose that G/K admits a G-equivariant spin^c structure. One forms the twisted Dirac operator $D_{\mathcal{V}}$ and by taking the higher index, the Dirac induction (3.5.2.1) factors through the following morphism:

$$\nu : K_G^0(G/K) \to K_m(C_r^*(G)) \qquad [\mathcal{V}] \mapsto [\mathrm{ind}_G(D_{\mathcal{V}})] \tag{3.5.20.2}$$

where m is the dimension of G/K. Then the Connes–Kasparov isomorphism is equivalent to ν being an isomorphism. Recall that since G/K carries a G-equivariant spin^c structure, one has the Poincaré duality

$$K_G^0(G/K) \cong K_m^G(G/K) \qquad [\mathcal{V}] \mapsto [D_{\mathcal{V}}] \tag{3.5.20.3}$$

where $m = \dim G/K$ and $K_m^G(G/K)$ is the equivariant K-homology of G/K, represented by abstract G-invariant elliptic operators on a Hilbert space, admitting a $C_0(G/K)$ representation through pointwise multiplications, in the form of a KK-cycle. Composing (3.5.20.2) with (3.5.20.3), one obtains a morphism given by the higher index

$$\mu : K_m^G(M) \to K_m(C_r^*(G)) \qquad [D_{\mathcal{V}}] \mapsto \mathrm{ind}_G(D_{\mathcal{V}}).$$

This map μ can be defined without assuming G/K having a G-equivariant spin^c structure and D_V can be replaced by an arbitrary G-invariant elliptic operator on M. By definition, μ is the Baum–Connes assembly map specialized to the case of a connected Lie group. Therefore, we see that the Baum–Connes conjecture for connected Lie groups is equivalent to the Connes–Kasparov conjecture.

The Baum–Connes conjecture is proved to be true for an almost connect Lie group [16, 34, 39] and hence the Connes–Kasparov conjecture is verified:

Theorem 3.5.21 (Connes–Kasparov isomorphism)**.** *When G is an almost connected Lie group, the Baum–Connes assembly map*

$$\mu : K_n^G(G/K) \to K_n(C_r^*(G)) \tag{3.5.21.1}$$

is an isomorphism. When G is connected, this gives rise to the isomorphism of the Dirac induction map

$$R_{\text{spin}}(K) \to K_n(C_r^*(G)) \qquad [V] \to [\text{ind}_G(D_V)]$$

where $n = \dim G/K$.

The name Connes–Kasparov conjecture appeared first in early works by Plymen, see [44] for example. Logically, it can be interpreted by a combination of Connes' Thom isomorphism for crossed product by $\mathbb{R}^n$ [20] and Kasparov's Dirac-dual Dirac method in showing the Novikov conjecture for almost connected Lie groups [33, 34]. Before finishing the section, let us explore the two aspects slightly.

3.5.22 Connes' Thom Isomorphism and Mackey Machine

First, let us recall Connes' version of the Thom isomorphism.

Theorem 3.5.23 ([20])**.** *Let A be a C^*-algebra carrying an action by $\mathbb{R}^n$. Then*

$$K_i(A \rtimes \mathbb{R}^n) \cong K_{i+n}(A).$$

Example 3.5.24. When the action is trivial, we have

$$A \rtimes \mathbb{R}^n \cong A \otimes C^*(\mathbb{R}^n) \cong A \otimes C_0(\widehat{\mathbb{R}^n}) = A \otimes C_0(\mathbb{R}^n)$$

and the Connes' Thom isomorphism theorem reduces to the Bott periodicity:

$$K_i(A \otimes C_0(\mathbb{R}^{2n})) \cong K_i(A).$$

Recall that the Bott map $K_0(A) \to K_0(A \otimes C_0(\mathbb{R}^2))$ is defined by tensoring the Bott element on $\mathbb{R}^2$:

$$\beta := \frac{1}{|z|^2+1}\begin{bmatrix} |z|^2 & \bar{z} \\ z & 1 \end{bmatrix} - \begin{bmatrix} 1 & 0 \\ 0 & 0 \end{bmatrix}.$$

For a connected Lie group G with a maximal compact subgroup K, observe that G/K is diffeomorphic to the Euclidean space $\mathbb{R}^n$ when $n = \dim G/K$. Motivated by Connes' Thom isomorphism, it seems reasonable to conjecture that there could be an isomorphism

$$K_i(A \rtimes_r G) \cong K_{i+n}(A \rtimes K). \tag{3.5.24.1}$$

Remark 3.5.25. *Note that when $A = \mathbb{C}$, this reduces to the Connes–Kasparov isomorphism. However, when $A = C_0(G/\Gamma)$ where Γ is a discrete subgroup of G, the situation could be more complicated. In this case, $A \rtimes G$ is Morita equivalent to $C_r^*(\Gamma)$. Note that K acts on G/Γ properly. Suppose in addition that K acts freely on G/Γ. This can be achieved by replacing Γ with a finite index subgroup so that $\Gamma\backslash G/K$ is a manifold. For example, Γ can be assumed to be torsion-free. Then $A \rtimes K$ is Morita equivalent to $C_0(\Gamma\backslash G/K)$. In view of the Poincare duality, the isomorphism (3.5.24.1)*

$$K_i(C_r^*(\Gamma)) \cong K^{i+n}(\Gamma\backslash G/K)$$

is then equivalent to the Baum–Connes conjecture for Γ, which is widely open. For example, this happens when $G = SL(n,\mathbb{R})$ for $n \geq 3$ and Γ is a torsion-free subgroup of $SL(n,\mathbb{Z})$ of finite index.

Denote by x_0 the class of the group identity of G in G/K. It turns out to be quite interesting to reformulate the conjectural isomorphism (3.5.24.1) by introducing the space

$$V = T_{x_0}(G/K) = \mathfrak{g}/\mathfrak{k}$$

equipped with the adjoint action by $K = G_{x_0}$ and conjecture

$$K_i(A \rtimes_r G) \cong K_i((A \otimes C_0(V)) \rtimes K). \tag{3.5.25.1}$$

The reason for the right-hand side of (3.5.25.1) replacing that of (3.5.24.1) is motivated by the K-equivariant Bott periodicity. We remark that the K-equivariant Bott periodicity requires some extra conditions regarding orientation, see [23], or one can replace $C_0(V)$ by such functions with coefficients in the Clifford algebra $Cl(V)$. See the paper by Higson, Kasparov and Trout. In particular, when $A = \mathbb{C}$, (3.5.25.1) reduces to the Connes–Kasparov isomorphism. In fact,

$$K_i(C_r^*(G)) \cong K_i(C_0(V) \rtimes K) \cong K_i^K(C_0(V)) \cong \begin{cases} R_{\mathrm{spin}}(K) & i-n \text{ even} \\ 0 & i-n \text{ odd} \end{cases}.$$

Recall that $R_{\mathrm{spin}}(K)$ is generated by representations of a specific double cover $\widetilde{K}$ of K that do not factor through K and the last isomorphism is due to the equivariant Bott periodicity (for compact groups).

By rewriting $C_0(V) \rtimes K$ as $C^*(V) \rtimes K$ via Fourier transform and then as the C^*-algebra $C^*(V \rtimes K)$ of the semidirect product $V \rtimes K$:

$$C_0(V) \rtimes K \cong C^*(V) \rtimes K = C^*(V \rtimes K),$$

the Connes–Kasparov isomorphism is then equivalent to

$$K_i(C_r^*(G)) \cong K_i(C^*(V \rtimes K)).$$

Definition 3.5.26. Let $G_0 = V \rtimes K$ be the semidirect product group whose multiplication is given by

$$(v_1, k_1) \cdot (v_2, k_2) = (Ad_{k_2^{-1}}(v_1) + v_2, k_1 k_2).$$

The group G_0 is called the *motion group.*

The alternative description of the Connes–Kasparov isomorphism

$$K_i(C^*(G_0)) \cong K_i(C_r^*(G)) \tag{3.5.26.1}$$

can be proved using the idea of deformations. Recall that the idea of a K-theoretic index is to apply the (generalized) Fredholm property and use the boundary maps in K-theory. An alternative idea comes from the point of view of deforming algebras. Recall that for a closed manifold, Atiyah–Singer [7] showed that the analytic index can be described by

$$K_0(C_0(TM)) \to \mathbb{Z} \qquad [\sigma_D] \mapsto \operatorname{ind}(D)$$

where D is an elliptic pseudodifferential operator on M. Replacing $\mathbb{Z}$ by the K-theory of the C^*-algebra $C^*(M \times M)$ of the pair groupoid, Connes [21] used the notion of tangent groupoid to formulate the analytic index map

$$K_0(C_0(TM)) \to K_0(C^*(M \times M)) \cong \mathbb{Z}$$

as a result of the deformation-to-the-normal-cone procedure. The same idea was also reflected in the higher index theory, see Yu's localization algebra in [53] for example. Relative to the setting in (3.5.26.1), the isomorphism is a result of a continuous family of C^*-algebras $A_t, 0 \leq t \leq 1$ where $A_t \cong C_r^*(G)$ when $t \neq 0$ and $A_0 \cong C_r^*(G_0)$, where the continuity at 0 is induced by deformation to the normal cone relative to $K \subset G$ or, from the index theory point of view, to the diagonal inclusion

$$G/K \to (G/K) \times_G (G/K).$$

See [31] for example a precise construction of the continuous family of algebras.

The alternative formulation (3.5.26.1) of the Connes–Kasparov isomorphism is also quite famous because of its relationship to the Mackey analogy.

The *Mackey machine* predicts a one-to-one correspondence between the unitary dual of G and that of G_0:

$$\widehat{G} \leftrightarrow \widehat{G_0}.$$

The matching of representations under Mackey correspondence is closely related to the description of the representation of a semisimple Lie group via $(\mathfrak{g}, K)$-modules. The Connes–Kasparov isomorphism (3.5.26.1) is a K-theoretic version of the Mackey correspondence. The feature of the Mackey correspondence in the context of C^*-algebras can be studied by applying non-commutative harmonic analysis to both $C_r^*(G)$ and $C_r^*(G_0)$. This is done in the complex semisimple case in a well-known paper by Higson:

Theorem 3.5.27 ([29]). *Let G be a complex semisimple Lie group. The Mackey machine induces an isomorphism on K-theory*

$$K_i(C_r^*(G_0)) \cong K_i(C_r^*(G))$$

and verifies the Connes–Kasparov conjecture.

Higson's work was generalized to more representation theoretic settings, most notibly by Afgoustidis to the more complicated case of real reductive Lie groups [1–3]. See also Subag [47] and Tan-Yao-Yu [48].

3.5.28 Dirac-dual Dirac Method

Let G be a connected Lie group and K a maximal compact subgroup. Kasparov introduced the KK-theory approach to formulate the Dirac induction by introducing a Dirac element

$$[D] \in KK_G^n(C_0(G/K), \mathbb{C})$$

constructed from the Dirac operator D on the vector bundle $G \times_K \Delta$ associated with a G-equivariant spin^c structure and a spin representation Δ of K. The Dirac induction can be reformulated as follows. See [33, 38].

Proposition 3.5.29. *The Dirac induction map $R(K) \to K_0(C_r^*(G))$ is obtained from taking KK-product with the Dirac element $j^G[D]$:*

$$R(K) \cong K_0(C_r^*(K)) \to K_n(C_r^*(G)) \qquad [E] \mapsto [E] \otimes_{C^*(K)} j^G[D].$$

Proof. Let E be a finite-dimensional representation of K. Let $\Gamma(G \times_K E)$ be the sections of the homogeneous vector bundle $G \times_K E \to G/K$. It represents an element

$$[\Gamma(G \times_K E)] \in KK^G(C_0(G/K), C_0(G/K)).$$

Then under the descend map one has

$$j^G : KK^G(C_0(G/K), C_0(G/K)) \to KK(C^*(K), C^*(K)) \quad [\Gamma(G \times_K E)] \to [E].$$

Here we used the fact that $C_0(G/K) \rtimes G$ and $C^*(K)$ are Morita equivalent.

Moreover, under the isomorphism $R(K) \cong K_0(C_0(G/K) \rtimes G)$, the trivial representation $[1] \in R(K)$ corresponds the canonical projection $[p] \in K_0(C_0(G/K) \rtimes G)$.

Thus, under the KK product

$$R(K) \times KK(C^*(K), C^*(K)) \times KK(C^*(K), C_r^*(G)) \to K_0(C_r^*(G))$$

one has

$$[1]_{C^*(K)} \otimes [E]_{C_0(G/K)\rtimes G}[j^G(D)]$$
$$= [p] \otimes_{C_0(G/K)\rtimes G} j^G([\Gamma(G \times_K E)]) \otimes_{C_0(G/K)\rtimes G} [j^G(D)].$$

Note that the left-hand side is reduced to $[E] \otimes_{C_0(G/K)\rtimes G} [j^G(D)]$ and the right-hand side is equal to

$$[p] \otimes_{C_0(G/K)\rtimes G} j^G(D_E) =: \mathrm{ind}_G(D_E).$$

Therefore $\mathrm{ind}_G(D_E)$ given by the Dirac induction from E is equal to $[E] \otimes_{C_0(G/K)\rtimes G} [j^G(D)]$. The proposition is then proved. □

The idea of Dirac-dual Dirac method for showing the Connes–Kasparov conjecture is to produce a Bott element, or a dual Dirac element, $\beta \in KK_G^n(\mathbb{C}, C_0(G/K))$ inverse to the Dirac element in the sense of taking KK product:

$$[D] \otimes \beta = 1 \qquad \beta \otimes [D] = 1.$$

Thus, tensoring $j^G(\beta) \in KK^n(C_r^*(G), C^*(K))$ gives rise to the inverse to the Dirac induction.

In general, the Dirac-dual Dirac method is an effective way of showing the Baum–Connes conjecture. Usually, the complication is to construct the Bott element β, so that

$$[D] \otimes \beta = 1.$$

This implies that

$$\gamma = \beta \otimes [D] \in KK_G(\mathbb{C}, \mathbb{C})$$

is an idempotent with respect to the ring structure given by the Kasparov product. The construction of β or the existence of the γ element implies the injectivity of the assembly map, known as the Novikov conjecture. See for example [34, 36]. If one can show further that $\gamma = 1$ in $KK_G(\mathbb{C}, \mathbb{C})$, then the Baum–Connes assembly map is an isomorphism.

However, showing $\gamma = 1 \in KK_G(\mathbb{C},\mathbb{C})$ is not reasonable for property T groups, which is the case for many almost connected Lie groups. This issue was solved by Lafforgue by introducing the more flexible Banach KK-theory and showing that $\gamma = 1 \in KK_G^{ban}(\mathbb{C},\mathbb{C})$ and that almost connected Lie groups have property RD. See [38, 39, 50].

3.6 Sketch of Proof via Noncommutative Fourier Analysis

In this last section of the chapter, we sketch a proof of the Connes–Kasparov conjecture using noncommutative Fourier analysis. This approach was initiated in the work of Penington and Plymen [44] under the setting of a complex semisimple Lie group, and we shall mainly focus on this setting. These examples illustrate elegant connections between the K-theory of group C^*-algebras and the representation theory of Lie groups.

Let G be a complex or real reductive Lie group. Denote by $\widehat{G}$ the space of irreducible unitary representations of G under Fell topology. Let $\widehat{G}_t \subset \widehat{G}$ be the topological subspace of tempered representations of G, that is, the representation occurred in the left regular representations of G in $L^2(G)$. The topology of $\widehat{G}_t$ is in general T_0 and becomes Hausdorff for complex semisimple Lie groups. The idea of the proof is the following. First, identify $C_r^*(G)$ with its Fourier or Gelfand transform, that is, a family of algebras of compact operators over the tempered dual $\widehat{G}_t$. See [17]. Then, for a representation V of K, form the twisted Dirac operator D_V and identify its image $\mathrm{ind}_G(D_V) \in K_*(C_r^*(G))$ under Dirac induction with the family index of some Dirac operators parametrized by $\widehat{G}_t$. Finally, for every $[V] \in \widehat{K}$, use the properties of Dirac operators to identify $\mathrm{ind}_G(D_V)$ with a component of $\widehat{G}_t$ and match generators on both sides of the Dirac induction map.

3.6.1 Fourier Transform and Decomposition of $C_r^*(G)$, Example of $\mathbb{R}^n$

In this subsection, we use Fourier transform or harmonic analysis on $\mathbb{R}^n$ to study the higher index

$$\mathrm{ind}_G(D) \in K_m(C_r^*(\mathbb{R}^n))$$

sketched in Example 3.4.23.

Recall that for $G = \mathbb{R}^n$, the unitary and tempered duals

$$\widehat{G} = \widehat{G}_r = \mathbb{R}^n$$

are known as the Pontryagin dual of G. Let D be the Dirac operator on $\mathbb{R}^n$ and we consider its higher index

$$\mathrm{ind}_{\mathbb{R}^n}(D) \in K_n(C_r^*(\mathbb{R}^n)).$$

For the commutative algebra $C_r^*(G)$, one has the Fourier transform, also known as Gelfand transform

$$C_r^*(G) \to C_0(\widehat{G})$$
$$f \mapsto \widehat{f} \qquad \widehat{f}(\chi) := \int_G f(g)\chi(g)dg$$

which gives rise to isomorphisms $C_r^*(\mathbb{R}^n) \cong C_0(\widehat{R^n}) = C_0(\mathbb{R}^n)$. Therefore,

$$K_n(C_r^*(\mathbb{R}^n)) \cong K_n(C_0(\mathbb{R}^n)) \cong \mathbb{Z}. \tag{3.6.1.1}$$

To analyze the higher index, let us complete $C_c(\mathbb{R}^n, S)$ into a Hilbert $C_r^*(\mathbb{R}^n)$-module $\mathcal{E}(\mathbb{R}^n, S)$ as in Definition 3.4.14. In this case, $\mathcal{E}(\mathbb{R}^n, S) = L^2(\mathbb{R}^n, S)$. Let us take the KK-cycle $(\mathcal{E}(\mathbb{R}^n, S), D)$ representing $\mathrm{ind}_{\mathbb{R}^n}(D)$ and study the transformation under the Fourier transform.

Proposition 3.6.2. *Let $D : C_c^\infty(\mathbb{R}^n, S) \to C_c^\infty(\mathbb{R}^n, S)$ be the Dirac operator on $\mathbb{R}^n$. The higher index*

$$\mathrm{ind}_{\mathbb{R}^n}(D) \in K_n(C_r^*(\mathbb{R}^n))$$

is the Bott generator of $K_n(C_0(\mathbb{R}^n))$ under the isomorphism (3.6.1.1).

Proof. Denote by $H_\pi = \mathbb{C}$ the representation space of $\pi \in \widehat{\mathbb{R}^n}$. The Hilbert $C_r^*(\mathbb{R}^n)$-module $\mathcal{E}(\mathbb{R}^n, S)$ in this case is simply $L^2(\mathbb{R}^n, S)$ because the space $\mathbb{R}^n$ is also the group. Under the Fourier transform, $\mathcal{E}(\mathbb{R}^n, S)$ is identified with

$$L^2(\widehat{\mathbb{R}^n}, \{H_\pi \otimes S\}_{\pi \in \widehat{\mathbb{R}^n}})$$

determined by

$$\widehat{f}(\rho) = \int_{\mathbb{R}^n} \rho(g) f(g) dg \in L^2(\widehat{\mathbb{R}^n}, S \otimes H_\rho)$$

for $f \in L^2(\mathbb{R}^n, S)$ and $\rho \in \widehat{\mathbb{R}^n}$. Properties of Fourier transform imply that

$$\widehat{Df}(\rho) = D_\rho \widehat{f}(\rho) \qquad \rho \in \widehat{\mathbb{R}^n}$$

where D_ρ becomes the multiplication by the symbol $\sigma_D = ic(\xi)$ of D on $L^2(\widehat{\mathbb{R}^n}, S \otimes H_\pi)$:

$$[D_\rho \widehat{f}](\xi) = ic(\xi)\widehat{f}(\xi).$$

Therefore, the Fourier transform of D on $\mathcal{E}(\mathbb{R}^n, S)$ is identified with $\{D_\pi\}_{\pi \in \widehat{\mathbb{R}^n}}$ on $L^2(\widehat{\mathbb{R}^n}, \{H_\pi \otimes S\}_{\pi \in \widehat{\mathbb{R}^n}})$ where

$$D_\pi(\lambda \otimes s) = \lambda \otimes c(\pi)s \qquad \lambda \in H_\pi, s \in S$$

is determined by the Clifford multiplication. Observe that this exactly gives rise to the Bott generator for $KK(\mathbb{C}, C_0(\widehat{\mathbb{R}^n}))$. Therefore, we have proved the proposition. □

In summary, we perform the Fourier transform of D on $\mathcal{E}(\mathbb{R}^n, S) = L^2(\mathbb{R}^n) \otimes \Delta$ so that it gives rise to a cycle $(L^2(\mathbb{R}^n, S), D)$ in $K_n(C_0(\mathbb{R}^n))$, and find the Fourier transform under the Plancherel decomposition:

$$L^2(\mathbb{R}^n) = \int_{\widehat{\mathbb{R}^n}} H_\pi \otimes H_\pi^* d\mu(\pi)$$

and D is decomposed into

$$D_\pi : H_\pi \otimes (H_\pi^* \otimes \Delta) \to H_\pi \otimes (H_\pi^* \otimes \Delta)$$

given by the Clifford multiplication by $ic(\pi)$ on Δ for every $\pi \in \widehat{\mathbb{R}^n}$.

3.6.3 Case of Complex Semisimple Lie Groups

We show the Connes–Kasparov isomorphism for a complex semisimple Lie group, following closely [44]. The basic idea was discussed in case of $\mathbb{R}^n$, but one needs finer results involving representations of G and Dirac operators on the associated homogeneous space G/K.

Let G be a connected complex semisimple group. We shall first review the structure of its reduced C^*-algebra. See [17] for details in a more general setting. Semisimple or reductive groups have type 1 C^*-algebras. This implies that for every $f \in C_c^\infty(G)$ and $(\pi, H_\pi) \in \widehat{G}$,

$$\widehat{f}(\pi) = \pi(f) := \int_G f(g)\pi(g)dg \in \mathcal{B}(H_\pi)$$

is a trace class operator on H_π. By the universal property of the maximal group C^*-algebra, the map π extends to a morphism of C^*-algebras

$$\pi : C^*(G) \to \mathcal{K}(H_\pi) \qquad \pi(f) = \widehat{f}(\pi). \tag{3.6.3.1}$$

For a tempered representation π of G, the map (3.6.3.1) factors through the reduced group C^*-algebra:

$$\pi : C_r^*(G) \to \mathcal{K}(H_\pi) \qquad \pi(f) = \widehat{f}(\pi).$$

It turns out that the tempered dual $\widehat{G}_t$ is a Hausdorff space for the complex semisimple Lie groups G. Furthermore, from the representation theory of semisimple Lie groups, points in $\widehat{G}_t$ can be identified with $(\widehat{M} \times \widehat{A})/W$ where M, A are in the Langlangds decomposition of the minimal parabolic subgroup and W is the Weyl group.

Recall that when $G = \mathbb{R}^n$, one has $C_r^*(G) \cong C_0(\widehat{G})$. In general, when G is complex semisimple or reductive, $C_r^*(G)$ can be identified with $C_0(\widehat{G}_t, \{\mathcal{K}(H_\pi)\}_{\pi \in \widehat{G}_t})$ or a subalgebra of $C_0(\widehat{G}_t) \otimes \mathcal{K}$.

Example 3.6.4. Let $G = SL(2, \mathbb{C})$ and $K \cong SU(2)$ be the maximal compact subgroup. Then $G = KP$ where

$$P = \left\{ \begin{bmatrix} a & b \\ 0 & a^{-1} \end{bmatrix} | a \in \mathbb{C} \backslash \{0\}, b \in \mathbb{C} \right\}$$

is the minimal parabolic subgroup. The group P has the Langlands decomposition $P = MAN$ where

$$M = \left\{ \begin{bmatrix} a & 0 \\ 0 & a^{-1} \end{bmatrix} | |a| = 1 \right\}, \quad A = \left\{ \begin{bmatrix} a & 0 \\ 0 & a^{-1} \end{bmatrix} | a > 0 \right\} \quad N = \left\{ \begin{bmatrix} 1 & b \\ 0 & 1 \end{bmatrix} | b \in \mathbb{C} \right\}.$$

Every tempered irreducible representation π of G can be obtained by the parabolic induction

$$\pi = \pi_{\sigma,\mu} = \mathrm{Ind}_{P=MAN}^{G} \sigma \otimes \mu \otimes 1$$

for some $(\sigma, \mu) \in \widehat{M} \times \widehat{A}$. Denote by $H_{\sigma,\mu}$ the representation space of $\pi_{\sigma,\mu}$. The Weyl group $W = \mathbb{Z}/2$ acts by conjugating $\begin{bmatrix} 0 & -1 \\ 1 & 0 \end{bmatrix}$ on M, A and it maps (σ, μ) to $(-\sigma, -\mu)$. The representations $\pi_{\sigma,\mu}$ and $\pi_{\sigma',\mu'}$ are equivalent if and only if $(\sigma, \mu) = -(\sigma', \mu')$. Let $[\sigma, \mu]$ be the equivalence class of $(\sigma, \mu) \in \widehat{M} \times \widehat{A}$ under this equivalence relation. Thus, one has the following precise identification:

$$(\widehat{M} \times \widehat{A})/W \to \widehat{G}_t \qquad [\sigma, \mu] \mapsto \pi_{\sigma,\mu}.$$

The structure of the reduced group C^*-algebra is given by

$$C_r^*(G) \cong C_0((\widehat{M} \otimes \widehat{A})/W) \otimes \mathcal{K} \cong C_0((\widehat{M} \otimes \widehat{A})/W, \{\mathcal{K}(H_{\sigma,\mu})\}_{(\sigma,\mu) \in \widehat{G}_t})$$

determined by the Fourier transform:

$$\widehat{f}([\sigma, \mu]) := \int_G f(g) \pi_{\sigma,\mu}(g) dg \in \mathcal{K}(H_{\sigma,\mu}) \qquad \forall f \in C_c(G) \subset C_r^*(G).$$

Next, we shall compute the higher index of the Dirac operator on G/K twisted by a representation of K. Let $V = V_\tau \in \widehat{K}$ where $\tau \in i\mathfrak{t}$ is the highest weight of V in the positive Weyl chamber of K. So $[V] \in R(K)$. Denote by

$$D_{V_\tau} : (L^2(G) \otimes \Delta \otimes V_\tau)^K \to (L^2(G) \otimes \Delta \otimes V_\tau)^K$$

the Dirac operator twisted by the homogenous vector bundle $G \times_K V_\tau$ over G/K. We shall compute the higher index

$$\mathrm{ind}_G(D_{V_\tau}) = [((L^2(G) \otimes \Delta \otimes V_\tau)^K, D_{V_\tau})] \in KK_n(\mathbb{C}, C_r^*(G))$$

where $n = \dim(G/K) = \dim A$, using the Fourier transform. This will lead to the main conclusion of this chapter:

Theorem 3.6.5 ([44]). *Let G be a connected complex semisimple Lie group and K be a maximal compact subgroup. Then the Dirac induction is an isomorphism*

$$R(K) \to K_n(C_r^*(G)) \qquad n = \dim(G/K)$$

sending $[V_\tau] \in \widehat{K}$ with the highest weight τ to the component for $\sigma \in \widehat{M}$ with the highest weight $\tau + \rho_K$, where ρ_K is the half sum of compact positive roots.

Recall that $\widehat{G}_t = (\widehat{M} \times \widehat{A})/W$. So for every $\sigma \in \widehat{M}$, there corresponds the component $(\{\sigma\} \times \widehat{A})/W_\sigma$ of $\widehat{G}_t$. Note also that generators of $K_*(C_r^*(G))$ are identified with components that are homeomorphic to $\widehat{A}$, hence labeled by $\sigma \in \widehat{M}$. To show the theorem, one needs to match every irreducible representation of K with a component associated with $\sigma \in \widehat{M}$. The higher index $\text{ind}_G(D_{V_\tau})$ will be proved to be the family index of a family of operators over a component, in terms of the Bott element on the component. To be more precise, the following steps are involved.

1. Decompose $C_r^*(G)$ using Fourier transformation;
2. Decompose $(L^2(G) \otimes \Delta \otimes V_\tau)^K$ using Plancherel formula for $L^2(G)$;
3. Find the Fourier transform of D at each $(\pi, H_\pi) \in \widehat{G_t}$ and denote by $D_\pi : (H_\pi \otimes \Delta \otimes V_\tau)^K \to (H_\pi \otimes \Delta \otimes V_\tau)^K$ the restriction of the Fourier transform of D at π;
4. Study properties of D_π and $(H_\pi \otimes \Delta \otimes V_\tau)^K$. In fact
   - D_π^2 is a scalar multiplication;
   - D_π does not contribute to the family index if D_π^2 is strictly positive;
   - $(H_\pi \otimes \Delta \otimes V_\tau)^K$ has finite dimension and vanishes on many occasions.
5. Conclude that the family $\{D_{\sigma,\lambda}\}_{\lambda \in \widehat{A}}$ is a Bott generator on the $\{\sigma\} \times \widehat{A}$ component labeled by $\sigma \in \widehat{M}$.

We sketch the proof of Theorem 3.6.5 following these steps.

Step 1 Decompose $C_r^*(G)$: Recall that under the Fourier transform,

$$C_r^*(G) \cong C_0(\widehat{G_t}, \mathcal{K})$$

for a complex semisimple Lie group. Denote the connected components of $\widehat{G_t} = (\widehat{M} \times \widehat{A})/\mathbb{Z}_2$ by

$$C_\sigma := \begin{cases} \{\sigma\} \times \widehat{A} & W_\sigma = 1 \\ \{\sigma\} \times (\widehat{A}/W_\sigma) & W_\sigma \neq 1 \end{cases}$$

Then

$$C_r^*(G) \cong \bigoplus_{\sigma \in \widehat{M}/W} C_0(C_\sigma, \mathcal{K}(H_\sigma)), \qquad H_\sigma = \mathrm{Ind}_P^G C_0(\widehat{A}, V_\sigma)$$

where V_σ is the representation space of $\sigma \in \widehat{M}$, $C_0(\widehat{A}, V_\sigma)$ is a representation space for $\widehat{A} \times \{\sigma\}$ and every tempered irreducible representation of G has the form $(\pi_{\sigma,\lambda}, H_{\sigma,\lambda}) = \mathrm{Ind}_P^G(\sigma \otimes \lambda \otimes 1)$ for some $\sigma \in \widehat{M}, \lambda \in \widehat{A}$ and

$$H_\sigma := \int_{C_\sigma}^{\oplus} H_{\sigma,\lambda} d\mu(\lambda).$$

Step 2 Decompose $(L^2(G) \otimes \Delta \otimes V_\tau)^K$:

Applying the Plancherel decomposition to $L^2(G)$, one has

$$\begin{aligned}(L^2(G) \otimes \Delta \otimes V_\tau)^K &= \bigoplus_{\sigma \in \widehat{M}/W} \int_{C_\sigma} (H_{\sigma,\lambda} \otimes H_{\sigma,\lambda}^* \otimes \Delta \otimes V_\tau)^K d\mu(\lambda) \\ &= \bigoplus_{\sigma \in \widehat{M}/W} \int_{C_\sigma} H_{\sigma,\lambda} \otimes (H_{\sigma,\lambda}^* \otimes \Delta \otimes V_\tau)^K d\mu(\lambda).\end{aligned} \tag{3.6.5.1}$$

A key observation here is that by Schur orthogonality, many of the summands vanish:

Lemma 3.6.6 (Lemma 5.3 of [44]).

$$\dim(H_{\sigma,\lambda}^* \otimes \Delta \otimes V_\tau)^K = \begin{cases} 2^{\lfloor \frac{1}{2}\dim A \rfloor} & \sigma = \tau + \rho_K \\ 0 & \tau + \rho_K - \sigma \notin \textit{positive Weyl chamber} \end{cases}$$

Proof. The idea is sketched as follows. Let $W_a \in \widehat{K}$ be with the highest weight $a \in it^*$. Denote by $W_{<a}$ the sum of irreducible K-modules with a as a weight but not the highest weight. Note that

$$H_{\sigma,\lambda}|_K = \mathrm{Ind}_P^G(\sigma \otimes \lambda \otimes 1)|_K = \mathrm{Ind}_M^K \sigma.$$

Then as K-modules, we have

$$H_{\sigma,\lambda}|_K = W_\sigma \oplus W_{<\sigma}.$$

Decomposing Δ and V_τ into weight spaces, one has

$$\Delta = 2^{\lfloor n/2 \rfloor} \mathbb{C}_{\rho_K} \oplus \text{lower weights} \qquad V_\tau = \mathbb{C}_\tau \oplus \text{lower weights}$$

where $n = \dim A$. Thus

$$\Delta \otimes V_\tau = 2^{\lfloor n/2 \rfloor} W_{\rho_K + \tau} \oplus W_{<\rho_K + \tau}.$$

Then the lemma follows as a result of Schur's lemma. □

Step 3 Fourier transform D_{V_τ}: For a Dirac operator D on $(L^2(G)\otimes\Delta\otimes V_\tau)^K$, denote by $D_{\sigma,\lambda}$ its restriction to the component $H_{\sigma,\lambda} \otimes (H^*_{\sigma,\lambda} \otimes \Delta \otimes V_\tau)^K$ in (3.6.5.1). Then, by the property of Fourier transform, one has:

$$\widehat{Df}(\pi_{\sigma,\lambda}) = D_{\sigma,\lambda}\widehat{f}(\pi_{\sigma,\lambda})$$

for $f \in (L^2(G) \otimes \Delta \otimes V_\tau)^K$ and $\pi_{\sigma,\lambda} \in \widehat{G_t}$. Under Fourier transform:

$$\mathrm{ind}_G(D_{V_\tau}) = [((L^2(G) \otimes \Delta \otimes V_\tau)^K, D_{V_\tau})] \in KK_n(\mathbb{C}, C^*_r(G))$$

is decomposed into a direct sum over $\widehat{M}/W$, such that over each $[\sigma] \in \widehat{M}/W$, $\mathrm{ind}_G(D_{V_\tau})$ is given by

$$\left[\left(\int_{C_\sigma} H_{\sigma,\lambda} \otimes (H^*_{\sigma,\lambda} \otimes \Delta \otimes V_\tau)^K d\mu(\lambda), \{(D_{V_\tau})_{\sigma,\lambda}\}_{\lambda\in\widehat{A}}\right)\right]$$

in $KK_n(\mathbb{C}, C_0(\widehat{A}, \mathcal{K}(H_\sigma)))$.

Step 4 Properties of $D_{\sigma,\lambda}$: Recall Parthasarathy's precise calculation of the square of the Dirac operator:

Lemma 3.6.7 ([43]). *For $D = D_{V_\tau}$, the square of $(D_{V_\tau})_{\sigma,\lambda}$ is a scalar given by*

$$(D_{V_\tau})^2_{\sigma,\lambda} = -\|\sigma\|^2 + \|\lambda\|^2 + \|\tau + \rho_K\|^2.$$

Here, $\|\cdot\|$ is defined by the inner product on $i\mathfrak{t}^*$ and $\sigma \in \widehat{M}$ is identified with its highest weight in $i\mathfrak{t}^*$.

Corollary 3.6.8. *If $\tau + \rho_K - \sigma$ is not zero and belongs to the positive Weyl chamber, then $(D_{V_\tau})^2_{\sigma,\lambda} > 0$ and hence $(D_{V_\tau})_{\sigma,\lambda}$ is invertible.*

Proof. By assumption, $\sigma = \tau + \rho_K - \Sigma$ where Σ is a positive combination of positive roots of K. Therefore

$$\begin{aligned}(D_{V_\tau})^2_{\sigma,\lambda} &= -\|\sigma\|^2 + \|\lambda\|^2 + \|\tau + \rho_K\|^2 \\ &= -\|\sigma\|^2 + \|\lambda\|^2 + \|\sigma + \Sigma\|^2 \geq -\|\sigma\|^2 + \|\sigma + \Sigma\|^2 > 0\end{aligned}$$

and $(D_{V_\tau})_{\sigma,\lambda}$ is invertible. □

From the two key properties, Lemma 3.6.6 and Corollary 3.6.8, only when $\sigma = \tau + \rho_K$ the component labeled by σ will survive in

$$\mathrm{ind}_G(D_{V_\tau}) \in KK(\mathbb{C}, C^*_r(G)) = \bigoplus_{[\sigma]\in\widehat{M}/W} KK_n(\mathbb{C}, C_0((\{\sigma\} \times \widehat{A})/W_\sigma)).$$

Note that the σ-component gives rise to a nontrivial generator when $W_\sigma = 1$ or when $C_\sigma = (\{\sigma\} \times \widehat{A})/W_\sigma \cong \widehat{A}$. This is summarized in the following.

Proposition 3.6.9. *The component for $\sigma \in \widehat{M}$ is nonzero in K-theory if and only if there exists $w \in W$ such that the highest weight of $w\sigma$ has the form $\tau + \rho_K$ for some τ in the positive Weyl chamber for K.*

Step 5 $\{D_{\sigma,\lambda}\}_{\lambda\in\widehat{A}}$ is the Bott generator: In view of Lemma 3.6.9,

$$\mathrm{ind}_G(D) = [((L^2(G) \otimes \Delta \otimes V_\tau)^K, D_{V_\tau})] \in KK_n(\mathbb{C}, C_r^*(G))$$

is reduced to

$$\left[\left(\int_{\widehat{A}} H_{\tau+\rho_K,\lambda} \otimes (H^*_{\tau+\rho_K,\lambda} \otimes \Delta \otimes V_\tau)^K d\mu(\lambda), \{(D_{V_\tau})_{\tau+\rho_K,\lambda}\}_{\lambda\in\widehat{A}}\right)\right]$$
$$\in KK_n(\mathbb{C}, C_0(\{\tau+\rho_K\} \times \widehat{A})).$$

Denote $\mathbb{S} = (H^*_{\tau+\rho_K,\lambda} \otimes \Delta \otimes V_\tau)^K$. It has $\dim 2^{\lfloor \dim A/2 \rfloor}$ by Lemma 3.6.6. Since $\mathrm{ind}_G(D)$ is a $C_r^*(G)$-module, the family operator

$$(D_{V_\tau})_{\tau+\rho_K,\lambda} \in \mathrm{End}(H_{\tau+\rho_K,\lambda} \otimes \mathbb{S})$$

is a $C_0(\widehat{A}, \mathcal{K}(H_{\tau+\rho_K}))$-module, or a $C_0(\widehat{A}, \mathcal{K}(H_{\tau+\rho_K}))$-linear map. That means that $(D_{V_\tau})_{\tau+\rho_K,\lambda}$ commutes with $\mathcal{K}(H_{\tau+\rho_K,\lambda})$. Thus, $(D_{V_\tau})_{\tau+\rho_K,\lambda}$ has the form

$$1 \otimes M_{\tau,\lambda} \qquad M_{\tau,\lambda} \in \mathrm{End}(\mathbb{S}).$$

Because $(D_{V_\tau})^2_{\tau+\rho_K,\lambda} = \|\lambda\|^2$, one has

$$(D_{V_\tau})_{\tau+\rho_K,\lambda} = 1 \otimes ic(\lambda).$$

Then, under Morita equivelence $C_0(\widehat{A}, \mathcal{K}(H_{\tau+\rho_K})) \sim C_0(\widehat{A})$, we have the Bott generator:

$$\left[\left(\int_{\widehat{A}} H_{\tau+\rho_K,\lambda} \otimes \mathbb{S}, 1 \otimes ic(\lambda)\right)\right] = \left[\left(\int_{\widehat{A}} \mathbb{S} d\mu(\lambda), ic(\lambda)\right)\right]$$

Thus, we have proved:

Proposition 3.6.10. *The higher index*

$$\mathrm{ind}_G(D) = [((L^2(G) \otimes \Delta \otimes V_\tau)^K, D_{V_\tau})] \in KK_n(\mathbb{C}, C_r^*(G))$$

is given by the Bott element associated with the component labeled by $\sigma = \tau + \rho_K$

$$\left[\left(\int_{\widehat{A}} \mathbb{S} d\mu(\lambda), ic(\lambda)\right)\right] \in KK_n(\mathbb{C}, C_0(\{\tau+\rho_K\} \times \widehat{A})).$$

Therefore, Theorem 3.6.5 is proved.

3.6.11 Example of $SL(2,\mathbb{R})$

In this subsection, we briefly discuss the example of Connes–Kasparov conjecture for $SL(2,\mathbb{R})$ over discrete series generators. For the more general setting of real reductive Lie groups, refer to [18, 19, 51]. Let us first recall the structure of the reduced algebra.

Theorem 3.6.12 ([51], [17]). *Let G be a connected real semisimple Lie group. The Fourier transform gives rise to an isomorphism*

$$C_r^*(G) \cong \bigoplus_{[(P,\sigma)]} C_0(\mathfrak{a}_P^*, \mathcal{K}(\mathrm{Ind}_P^G(H_\sigma)))^{W_\sigma}.$$

The discrete series components in the tempered dual come from $P = G$ and labeled by $\sigma \in \widehat{G_d}$. The corresponding C^*-component

$$C_0(\mathfrak{a}_P^*, \mathcal{K}(\mathrm{Ind}_P^G(H_\sigma)))^{W_\sigma}$$

reduces to $\mathcal{K}(H_\sigma)$ and

$$\mathbb{Z} \cong K_0(\mathcal{K}(H_\sigma)) \leq K_0(C_r^*(G)).$$

In the example of $SL(2,\mathbb{R})$, the discrete series components give rise to all but one K-theory generators:

$$K_0(C_r^*(SL(2,\mathbb{R}))) \cong K_0(C_0(\mathbb{R}) \rtimes \mathbb{Z}_2) \bigoplus \left(\bigoplus_{n\neq 0} K_0(\mathcal{K}(H_n)) \right).$$

Here, $H_n := D_n^+$ if $n > 0$ and $H_n := D_n^-$ if $n < 0$. Let $G = SL(2,\mathbb{R})$ and $K = SO(2)$. Let V_k be irreducible representation of K with weight $k \in \mathbb{Z}$. Consider the Dirac operator

$$D_{V_k} : (L^2(G) \otimes \Delta \otimes V_k)^K \to (L^2(G) \otimes \Delta \otimes V_k)^K.$$

When $k \neq 0$, as in the complex semisimple case, the principle series component of the image of the Dirac induction

$$\mathrm{ind}_G(D_{V_k}) = [((L^2(G) \otimes \Delta \otimes V_k)^K, D_{V_k})] \in KK(\mathbb{C}, C_r^*(G))$$
$$K_0(C_r^*(G)) \cong K_0(C_0(\mathbb{R}) \rtimes \mathbb{Z}_2) \bigoplus \left(\bigoplus_{n\neq 0} K_0(\mathcal{K}(H_n)) \right)$$

vanishes. Under the Fourier transform, $\mathrm{ind}_G(D_{V_k}) \in K_0(C_r^*(G))$ reduces to

$$\bigoplus_{n\neq 0} (H_n \otimes (H_n^* \otimes \Delta \otimes V_k)^K, (D_{V_k})_n) \in \bigoplus_{n\neq 0} K_0(\mathcal{K}(H_n)).$$

Under Morita equivalence, this is further simplified to

$$\bigoplus_{n\neq 0}((H_n^* \otimes \Delta \otimes V_k)^K, M_{k,n}) \in \bigoplus_{n\neq 0} K_0(\mathbb{C}).$$

Because $(H_n^* \otimes \Delta \otimes V_k)^K$ has finite dimension and $\mathbb{Z}_2$-graded, the image in $K_0(\mathbb{C}) \cong \mathbb{Z}$ can be calculated by the index

$$\bigoplus_{n\neq 0}[\dim(H_n^* \otimes \Delta^+ \otimes V_k)^K - \dim(H_n^* \otimes \Delta^- \otimes V_k)^K] \in \bigoplus_{n\neq 0} \mathbb{Z}$$

By Atiyah–Schmid [5], the number

$$\dim(H_n^* \otimes \Delta \otimes V_k)^K - \dim(H_n^* \otimes V^- \otimes V_k)^K = \begin{cases} 1 & n = k \\ 0 & n \neq k \end{cases}$$

So

$$\mathrm{ind}_G(D_{V_k}) = [H_k] \in K_0(C_r^*(G)) \qquad k \neq 0.$$

The identification of $[V_0] \in R(K)$ with the principal series component generator $K_0(C_0(\mathbb{R}) \rtimes \mathbb{Z}_2)$ by Fourier transform involves subtle harmonic analysis and representation theoretic techniques. See [15] and [18, 19] for full details.

References

[1] A. Afgoustidis. *On the analogy between real reductive groups and Cartan motion groups: a proof of the Connes-Kasparov isomorphism*. J. Funct. Anal. 277 (7) (2019), 2237–2258.

[2] A. Afgoustidis. *On the analogy between real reductive groups and Cartan motion groups: contraction of irreducible tempered representations*. Duke Math. J. 169 (5) (2020), 897–960.

[3] A. Afgoustidis. *On the analogy between real reductive groups and Cartan motion groups: the Mackey-Higson bijection*. Camb. J. Math. 9 (3) (2021), 551–575.

[4] M. F. Atiyah and R. Bott, *A Lefschetz fixed point formula for elliptic complexes. II. Applications*. Ann. Math. 88 (2) (1968), 451–491.

[5] M. Atiyah and W. Schmid. *A geometric construction of the discrete series for semisimple Lie groups*. Invent. Math. 42 (1977), 1–62.

[6] M. F. Atiyah and G. B. Segal. *The index of elliptic operators. II*. Ann. Math. 87 (2) (1968), 531–545.

[7] M. F. Atiyah and I. M. Singer. *The index of elliptic operators. I*. Ann. Math. 87 (2) (1968), 484–530.

[8] P. Baum and A. Connes. *Geometric K-theory for Lie groups and foliations*. Enseign. Math. 46 (2) (2000), 1–2, 3–42.

[9] P. Baum, A. Connes, and N. Higson. Classifying space for proper actions and K-theory of group C^*-algebras. In: *C^*-algebras: 1943–1993 A Fifty Year Celebration* (ed. R. Doran, San Antonio, TX, 1993), 241–291, Contemporary Mathematics, 167, American Mathematical Society, Providence, RI, 1994.

[10] P. Baum and R. G. Douglas. K homology and index theory. In: *Operator algebras and applications*, Part 1 (Kingston, ON, 1980), 117–173, Proceedings of Symposia in Pure Mathematics, 38, American Mathematical Society, Providence, R.I., 1982.

[11] S. Baaj and P. Julg. *Théorie bivariante de Kasparov et opérateurs non bornés dans les C^*-modules hilbertiens*. C. R. Acad. Sci. Paris. I Math. 296 (21) (1983), 875–878.

[12] N. Berline, E. Getzler, and M. Vergne. *Heat kernels and Dirac operators*. Corrected reprint of the 1992 original. Grundlehren Text Editions. Springer-Verlag, Berlin, 2004. x+363 pp.

[13] R. Bott. The index theorem for homogeneous differential operators. In: *Differential and combinatorial topology (A symposium in honor of Marston Morse)*, 167–186, Princeton University Press, Princeton, NJ,1965.

[14] R. Bott. *Homogeneous vector bundles*. Ann. Math. 66 (2) (1957), 203–248.

[15] J. Brodzki, G. A. Niblo, R. Plymen, and N. Wright. *The local spectrum of the Dirac operator for the universal cover of $SL_2(\mathbb{R})$*. J. Funct. Anal. 270 (3) (2016), 957–975.

[16] J. Chabert, S. Echterhoff, and R. Nest. *The Connes-Kasparov conjecture for almost connected groups and for linear p-adic groups*. Publ. Math. Inst. Hautes Études Sci. No. 97 (2003), 239–278.

[17] P. Clare, T. Crisp, and N. Higson. *Parabolic induction and restriction via C^*-algebras and Hilbert C^*-modules*. Compos. Math. 152 (6)(2016), 1286–1318.

[18] P. Clare, N. Higson, X. Tang, and Y. Song. *On the Connes-Kasparov isomorphism. I. The reduced C^*-algebra of a real reductive group and the K-theory of the tempered dual* . ArXiv: 2202.02855.

[19] P. Clare, N. Higson, and Y. Song. *On the Connes-Kasparov isomorphism. II. The Vogan classification of essential components in the tempered dual* . ArXiv: 2202.02857.

[20] A. Connes. *An analogue of the Thom isomorphism for crossed products of a C^*-algebra by an action of $\mathbb{R}$*. Adv. in Math. 39 (1) (1981), 31–55.

[21] A. Connes. *Noncommutative geometry*. Academic Press, Inc., San Diego, CA, 1994. xiv+661 pp.

[22] A. Connes and H. Moscovici. *The L^2-index theorem for homogeneous spaces of Lie groups*. Ann. of Math. 115 (2) (1982), 291–330.

[23] S. Echterhoff and O. Pfante. *Equivariant K-theory of finite dimensional real vector spaces*, Münster J. Math. 2 (2009), 65–94.

[24] J. M. G. Fell. *The dual spaces of C^*-algebras*, Trans. Amer. Math. Soc. 94 (1960), 365–403.

[25] H. Guo, P. Hochs, and V. Mathai. *Equivariant Callias index theory via coarse geometry*. Ann. Inst. Fourier (Grenoble) 71 (6) (2021), 2387–2430.

[26] Harish-Chandra. *Discrete series for semisimple Lie groups. I. Construction of invariant eigendistributions*. Acta Math. 113 (1965), 241–318.

[27] Harish-Chandra. *Discrete series for semisimple Lie groups. II. Explicit determination of the characters*. Acta Math. 116 (1966), 1–111.

[28] P. Hochs and H. Wang. *A fixed point formula and Harish-Chandra's character formula*. Proc. Lond. Math. Soc. 116 (1) (2018), 1–32.

[29] N. Higson. *The Mackey analogy and K-theory. Group representations, ergodic theory, and mathematical physics: A tribute to George W. Mackey*, 149–172, Contemp. Math., 2008.

[30] N. Higson and J. Roe. *Analytic K-homology*. Oxford Mathematical Monographs. Oxford University Press, 2000. xviii+405 pp.

[31] N. Higson and A. Román. *The Mackey bijection for complex reductive groups and continuous fields of reduced group C^*-algebras*. Represent. Theory 24 (2020), 580–602.

[32] J. S. Huang and P. Pandzic. *Dirac operators in representation theory*. Mathematics: Theory & Applications. Birkhäuser Boston, Inc., Boston, MA, 2006.

[33] G. G. Kasparov. *Operator K-theory and its applications: Elliptic operators, group representations, higher signatures, C^*-extensions*. Proceedings of ICM, Warsaw, 987–1000, 1984.

[34] G. G. Kasparov. *Equivariant KK-theory and the Novikov conjecture*. Invent. Math. 91 (1) (1988), 147–201.

[35] G. G. Kasparov. *Elliptic and transversally elliptic index theory from the viewpoint of KK-theory*. J. Noncommut. Geom. 10 (4) (2016), 1303–1378.

[36] G. Kasparov and G. Skandalis. *Groups acting properly on "bolic" spaces and the Novikov conjecture*. Ann. Math. 158 (1) (2003), 165–206.

[37] D. Kucerovsky. *A lifting theorem giving an isomorphism of KK–products in bounded and unbounded KK-theory*, J. Oper. Theory 44 (2000), 255–275.

[38] V. Lafforgue. *Banach KK-theory and the Baum-Connes conjecture*. Proceedings of the International Congress of Mathematicians, vol. II (Beijing, 2002), 795–812, Higher Education Press, Beijing, 2002.

[39] V. Lafforgue. *K-théorie bivariante pour les algébres de Banach et conjecture de Baum-Connes*. Invent. Math. 149 (1) (2002), 1–95.

[40] H. B. Jr. Lawson and M.-L. Michelsohn. *Spin geometry*. Princeton Mathematical Series, 38. Princeton University Press, Princeton, NJ, 1989. xii+427 pp.

[41] S. Mendes and R. J. Plymen. *Functoriality and K-theory for $GL_n(\mathbb{R})$*. Münster J. Math. 10 (2017), 39–58.

[42] B. Mesland and A. Rennie. *Nonunital spectral triples and metric completeness in unbounded KK-theory*, J. Funct. Anal. 271 (2016), 2460–2538.

[43] R. Parthasarathy. *Dirac operator and the discrete series*. Ann. Math. 96 (2) (1972), 1–30.

[44] M. G. Penington and R. J. Plymen. *The Dirac operator and the principal series for complex semisimple Lie groups*. J. Funct. Anal. 53 (3) (1983), 269–286.

[45] J. Roe. *Elliptic operators, topology and asymptotic methods*. Second edition. Pitman Research Notes in Mathematics Series, 395. Longman, Harlow, 1998. ii+209 pp.

[46] W. Schmid. *On a conjecture of Langlands*. Ann. Math. 93 (2) (1971), 1–42.

[47] E. Subag. *The algebraic Mackey-Higson bijections*. J. Lie Theory 29 (2) (2019), 473–492.

[48] Q. Tan, Y.-J. Yao, and S. Yu. *Mackey analogy via D-modules for $SL(2,\mathbb{R})$*. Internat. J. Math. 28 (7) (2017), 1750055, 20.

[49] A. Valette. *Dirac induction for semisimple Lie groups having one conjugacy class of Cartan subgroups*. 526–555, Lecture Notes in Math., 1132, Springer, 1985.

[50] A. Valette. *Introduction to the Baum-Connes conjecture*. From notes taken by Indira Chatterji. With an appendix by Guido Mislin. Lectures in Mathematics ETH Zürich. Birkhäuser Verlag, Basel, 2002. x+104 pp.

[51] A. Wassermann. *Une démonstration de la conjecture de Connes-Kasparov pour les groupes de Lie linéaires connexes réductifs*. C. R. Acad. Sci. Paris Sér. I Math. 304 (18) (1987), 559–562.

[52] N. E. Wegge-Olsen. *K-theory and C^*-algebras. A friendly approach.* Oxford Science Publications. The Clarendon Press, Oxford University Press, New York, 1993.

[53] R. Willett and G. Yu. *Higher index theory*. Cambridge Studies in Advanced Mathematics, 189. Cambridge University Press, Cambridge, 2020.

4

Representation Theory of p-Adic Reductive Groups

Anne-Marie Aubert

4.1 Introduction

This chapter is aimed at presenting some basic aspects of the representation theory of p-adic reductive groups (the reader will find more results on them in [22–24, 30, 34, 35, 72]), and mapping out some recent developments. This theory plays important roles in applications to automorphic forms and harmonic analysis – areas of central importance in the Langlands program.

A very useful tool to study representations of a given p-adic reductive group is provided by induction from proper subgroups. The very rich structure of p-adic reductive groups, which comes on the one hand from the general theory of reductive algebraic groups and on the other hand from the theory of Bruhat–Tits, is reflected in the existence of two types of induction: parabolic induction and compact induction. Parabolic induction is parallel to what one has for real Lie groups, while compact induction is special for the p-adic setting. The construction and classification of supercuspidal representations is a central problem in the subject, and key ingredients are provided by Bruhat–Tits buildings via the Moy–Prasad filtrations of parahoric subgroups and by the representation theory of finite reductive groups.

Particularly important results are the Bernstein decomposition of the category $\mathfrak{R}(G)$ of smooth representations of a p-adic reductive group G and the description of the structure of the set of isomorphism classes of irreducible smooth representations of G. The Bernstein blocks of $\mathfrak{R}(G)$ are equivalent to the module categories of certain generalized affine Hecke algebras, and studying the representation theory of these algebras appears to be a useful technique in understanding representations of G.

Another basic question is the harmonic analysis of representations of G, with the understanding of the Harish-Chandra characters of admissible representations.

Several other classes of representations of G are of fundamental interest: discrete series, tempered representations, and more generally unitary representations. The basic example of a discrete series representation that is (generally) not supercuspidal is the Steinberg representation. The Plancherel theorem is a huge generalization of the classical Fourier theory, in particular the Fourier inversion theorem/Plancherel formula: It expresses a sufficiently nice (e.g., Schwartz) function in terms of its "Fourier coefficients." It turns out that precisely the tempered representations contribute to the decomposition.

4.2 Smooth Representations of Locally Profinite Groups

4.2.1 Totally Disconnected Spaces

A Hausdorff locally compact topological space X is said to be totally disconnected if every point x of X admits a base of neighborhoods consisting of compact open subsets.

As a Hausdorff space, any two distinct points x and y of X belong to disjoint open neighborhoods. As a totally disconnected space, the disjoint neighborhoods of x and y can be made both compact and open. In this sense, the space X is totally disconnected for the only subsets of X that cannot be divided into a disjoint union of closed and open nonempty subsets are singletons. Nevertheless, in contrast with discrete sets, in general, singletons are not open subsets in a totally disconnected space.

Remark 4.2.1. A compact totally disconnected space X is a profinite set, that is, a limit of a projective system of finite sets equipped with the projective limit topology. Indeed, X being the union of its compact open subsets, which are themselves totally disconnected spaces, can be subdivided into a finite disjoint union of compact open subsets, which can be as small as desired.

4.2.2 Locally Profinite Groups

A topological group G is said to be profinite if it is compact and totally disconnected (i.e., any connected subset of G is a singleton). A finite group, with the discrete topology, is profinite. If G is profinite, then there is a topological isomorphism $G \simeq \varprojlim G/N$, where N runs through all open, normal subgroups of G, hence G is the projective limit of a system of finite groups. Conversely, such a projective limit is a compact and totally disconnected group, hence profinite.

Definition 4.2.2. A topological group G is called locally profinite if it satisfies one of the following equivalent conditions:

1. G is locally compact and totally disconnected;
2. every neighborhood of the identity in G contains a compact open subgroup.

Example 4.2.3. Compact locally profinite groups G are profinite.

4.2.3 Representations

A (complex) representation of a group G is a pair (π, V), where V is a complex vector space and π is a homomorphism from G to $\mathrm{Aut}_{\mathbb{C}}(V)$.

Definition 4.2.4. A representation (π, V) of G is *irreducible* if it is nonzero and the only G-subrepresentations of V are the trivial G-representation and V.

Let G be a locally profinite group.

Definition 4.2.5. A representation (π, V_π) of G is *smooth* if for any $v \in V_\pi$, its stabilizer

$$G_v := \{g \in G : \pi(g)(v) = v\}$$

in G is an open subgroup of G.

Let (π, V) be a representation of G. If K is a compact open subgroup of G, we write

$$V^K := \{v \in V : \pi(k)v = v \text{ for all } k \in K\}. \tag{4.2.3.1}$$

Remark 4.2.6. The representation (π, V) is smooth if and only if each vector v of V belongs to V^{K_v} for some compact open subgroup K_v of G.

The dual space V^* of V is equipped with the representation π^* of G defined by

$$\pi^*(g) \cdot \lambda(v) := \lambda(\pi(g^{-1}) \cdot v), \quad \text{for } \lambda \in V^* \text{ and } v \in V. \tag{4.2.3.2}$$

The representation (π^*, V^*) is not necessarily smooth. Let $\widetilde{V}$ denote the smooth part of V^*, it is stable under the action of G. We denote by $(\widetilde{\pi}, \widetilde{V})$ the restriction of π^* to $\widetilde{V}$: It is a smooth representation of G, called the *contragredient representation* of (π, V).

Smoothness is preserved by surjective morphisms and by the operation of taking subrepresentations, subquotients, and direct sums.

Proposition 4.2.7. If (π, V) is an irreducible smooth representation of G, then the dimension of V is countable.

Proof. Let v be a nonzero vector in V. Since (π, V) is smooth, there exists a compact open subgroup K of G such that $v \in V^K$. Since (π, V) is irreducible, $G \cdot v = G/K \cdot v$ spans the space V, and G/K is countable. □

We define

$$V^\infty := \{v \in V \,:\, G_v \text{ is open}\}. \tag{4.2.3.3}$$

Proposition 4.2.8. If (π, V) is a representation of G, then V^∞ is a smooth subrepresentation of V.

Proof. The set V^∞ is obviously invariant under scalar multiplication. Let $v, v' \in V^\infty$. Then there exists a compact open subgroup K of G that is contained in $G_v \cap G_{v'}$. Let $g \in G_{v+v'}$. Then gK is an open neighborhood of g, which is contained in $G_{v+v'}$. Thus, $G_{v+v'}$ is open. That is, $v + v' \in V^\infty$. It follows that V^∞ is a subspace of V. For $g \in G$ and $v \in V$, we have $G_{\pi(g)v} = gG_v g^{-1}$. Hence V^∞ is a subrepresentation of V, and it is smooth by definition. □

Definition 4.2.9. (Induced representations) Let H be a closed subgroup of G and let (τ, V_τ) be a smooth representation of H. Let $\mathrm{Ind}_H^G(V_\tau)$ denote the space of functions $f\colon G \to V_\tau$ such that

(1) for all $h \in H$ and $g \in G$, we have $f(hg) = \tau(h)f(g)$;
(2) there exists an open subgroup K_f of G such that for all $k \in K_f$ and $g \in G$, we have $f(gk) = f(g)$.

The space $\mathrm{Ind}_H^G(V_\tau)$ is stable under right translations by elements of G and condition (2) is the required smoothness condition, hence the representation $(\mathrm{Ind}_H^G \tau, \mathrm{Ind}_H^G V_\tau)$ of G is smooth and called the representation (smoothly) induced by τ.

The subspace $\mathrm{c{-}Ind}_H^G(V_\tau)$ of functions with compact support modulo H (i.e., the support is contained in some $H\Omega$ where Ω is a compact set in G) is G-stable and provides a subrepresentation $(\mathrm{c{-}Ind}_H^G(\tau), \mathrm{c{-}Ind}_H^G(V_\tau))$ of G called the representation compactly induced by τ. The two representations coincide when the quotient space G/H is compact.

4.2.4 Hecke Algebras

Let G be a locally profinite group. Let $C_c^\infty(G)$ denote the space of locally constant, compactly supported functions on G. A measure on G is a linear form on $\mathscr{C}_c^\infty(G)$. A (left) Haar measure on G is a nonzero measure m, which is invariant under left translation by G.

We fix a left Haar measure m on G. Then we define a product $*$, called convolution, on $\mathscr{C}_c^\infty(G)$ by setting

$$f_1 * f_2(x) := \int_G f_1(g) f_2(g^{-1}x) dm(g), \quad \text{for } f_1, f_2 \in \mathscr{C}_c^\infty(G) \text{ and } x \in G. \tag{4.2.4.1}$$

This product gives $\mathscr{C}_c^\infty(G)$ the structure of a $\mathbb{C}$-algebra, called the *Hecke algebra* of G and denoted $\mathscr{H}(G)$. This algebra has no unit element unless G is discrete.

4.3 Structure of p-Adic Reductive Groups

4.3.1 Non-Archimedean Local Fields

In the same way that the real numbers are typically constructed as the completion of the rationals with respect to the usual, intuitive Euclidean metric, we will construct the field $\mathbb{Q}_p$ of *p-adic numbers* by taking the completion of $\mathbb{Q}$ with respect to a metric induced by a more exotic norm, which we will denote $|\cdot|_p$, and refer to as the *p-adic norm* on $\mathbb{Q}$, defined as follows:

$$\left|\frac{a}{b}\right|_p := p^{v_p(b)-v_p(a)}, \tag{4.3.1.1}$$

where $v_p(n)$ is the exponent of p in the factorization of the natural integer n as a product of prime numbers. The map $x \mapsto |x|_p$ defines a homomorphism of abelian groups $\mathbb{Q}^\times \to \mathbb{R}^\times$. We extend it to a function $|\ \ |_p : \mathbb{Q} \to \mathbb{R}_{\geq 0}$ by setting $|0|_p := 0$. The p-adic absolute value can be extended by continuity to $\mathbb{Q}_p$ in a unique manner. Every element of $\mathbb{Q}_p$ can be represented by a Laurent series that converges under $|\ \ |_p$, that is, every element x of $\mathbb{Q}_p$ can be written in the form

$$x = a_{-m}p^{-m} + a_{-m+1}p_{-m+1} + \cdots + a_{-1}p_{-1} + a_0 + a_1^p + a_2^{p^2} + a_3^{p^3} + \cdots,$$

where $m \geq 0$ and $0 \leq a_i \leq p-1$ for all i. This is called the p-adic expansion of x. The ring of integers of $\mathbb{Q}_p$ is the ring of p-adic integers, defined as

$$\mathbb{Z}_p := \left\{x \in \mathbb{Q}_p \ : \ |x|_p \leq 1\right\}, \tag{4.3.1.2}$$

it consists precisely of those series such that $m = 0$, that is, the integers in $\mathbb{Q}_p$ have no negative powers of p in their p-adic expansion.

Let K be a field. A discrete valuation on K is a group homomorphism $v\colon K^\times \to \mathbb{Z}$ such that

$$v(x+y) \geq \min\{v(x), v(y)\} \quad \text{for all } x, y \in K^\times. \tag{4.3.1.3}$$

We extend the valuation to 0 by setting $v(0) = \infty$. With each valuation v on K, we associate a ring, called the *valuation ring of* v, given by

$$\mathfrak{o}_v := \{x \in K : v(x) \geq 0\} \tag{4.3.1.4}$$

It is easy to show that $\mathfrak{o}_v$ is an integral domain with quotient field K. In fact, $\mathfrak{o}_v$ is a local ring with unique maximal ideal, called the *valuation ideal*,

$$\mathfrak{p}_v := \{x \in K : v(x) > 0\}. \tag{4.3.1.5}$$

Each valuation induces an absolute value on K. Let $0 < \lambda < 1$ and define $|\cdot|_v : K \to \mathbb{R}$ by $|x|_v := \lambda v(x)$. Note that since valuations are additive, these absolute values are multiplicative. We say that two absolute values are equivalent if they induce the same topology on K. It follows that two absolute values $|\cdot|, |\cdot|'$ are equivalent if and only if there is some $s \in \mathbb{R}_+$ such that $|\cdot|' = |\cdot|^s$. Therefore, the value of λ chosen above does not change the equivalence class of the absolute value.

Let F be a *non-archimedean local field*, that is, F is complete with respect to the topology induced by a discrete valuation v and its *residue field* $k_F := \mathfrak{o}_v/\mathfrak{p}_v$ is finite. Then F is either a finite extension of $\mathbb{Q}_p$ or a field of formal Laurent series $\mathbb{F}_q((t))$, where $\mathbb{F}_q$ is a finite field with q elements and t is an indeterminate. The ring $\mathfrak{o}_F := \mathfrak{o}_v$ is called the *ring of integers of* F. We write $\mathfrak{p}_F := \mathfrak{p}_v$ for the maximal ideal in $\mathfrak{o}_F$, and we denote by $q = q_F$ the cardinality of the finite field k_F. We have $\mathfrak{p}_F = \varpi_F \mathfrak{o}_F$, the element ϖ_F is called a *uniformizer* of F.

Remark 4.3.1. The additive group $\mathfrak{o}_F$ of is a locally profinite group, with a fundamental system of neighborhoods of 0 given by $\mathfrak{p}^n$, $n \in \mathbb{Z}$. We have $\mathfrak{o}_F \simeq \varprojlim \mathfrak{o}_F/\mathfrak{p}^n$ for $n \geq 1$.

We write

$$U(\mathfrak{o}_F) = U^0(\mathfrak{o}_F) := \mathfrak{o}_F^\times \quad \text{and} \quad U^n(\mathfrak{o}_F) := 1 + \mathfrak{p}_F^n \quad \text{for } n \geq 1. \tag{4.3.1.6}$$

The group $U^n(\mathfrak{o}_F)$ is called the n-th principal congruence subgroup of $U(\mathfrak{o}_F)$.

4.3.2 The General Linear Group

Let N be a positive integer. Taking the topology on F determined by the absolute value $|\ |_F$, the set $\mathrm{M}_N(F) \simeq F^{N^2}$ of $N \times N$-matrices with entries in F is given the product topology.

The determinant $\det : \mathrm{M}_N(F) \to F$ is a polynomial in the matrix entries, hence it is a continuous map. The *general linear group* $\mathrm{GL}_N(F)$ is defined to be the inverse image of $F^\times = F\backslash\{0\}$ (an open subset of F) under the map det. It is an open subset of $\mathrm{M}_N(F)$ and we give it the topology it inherits in this way.

$$K_0 := \mathrm{GL}_N(\mathfrak{o}_F) = \{g \in \mathrm{GL}_N(F) \;:\; g \in \mathrm{M}_N(\mathfrak{o}_F) \text{ and } |\det(g)|_F = 1\}. \tag{4.3.2.1}$$

It is also an open compact subgroup of $\mathrm{GL}_N(F)$ and a maximal compact subgroup of $\mathrm{GL}_N(F)$.

Given $r \in \mathbb{R}_{>0}$, let $\mathrm{M}_N(\mathfrak{p}_F^{\lceil r\rceil})$ denote the subring of $\mathrm{M}_N(F)$ consisting of matrices the entries of which belong to $\mathfrak{p}^{\lceil r\rceil}$. Let

$$K_r := \left\{g \in \mathrm{GL}_N(F) \;:\; g - 1 \in \mathrm{M}_N(\mathfrak{p}_F^{\lceil r\rceil})\right\}. \tag{4.3.2.2}$$

Then K_r is an open compact subgroup of $\mathrm{GL}_N(F)$. The groups K_r give a filtration of K_0. For $r \geq 0$, let

$$K_{r+} := \bigcup_{r>s} K_s = 1 + \mathrm{M}_N(\mathfrak{p}_F^{\lceil r\rceil+1}). \tag{4.3.2.3}$$

Properties 4.3.2.

1. The K_r form a fundamental system of neighborhoods of 1; $\bigcap_{r>0} K_r = \{1\}$;
2. For $r, s \in \mathbb{R}_{>0}$, let $[K_r, K_s]$ denote the commutator group of K_r and K_s. We have $[K_r, K_s] \subset K_{r+s}$.
3. $\{r \in \mathbb{R}_0 \;:\; K_r \neq K_{r+}\}$ is discrete in $\mathbb{R}_{r>0}$.

We have an exact sequence

$$1 \longrightarrow K_{0+} \longrightarrow K_0 \xrightarrow{\mathrm{pr}} \mathrm{GL}_N(k_F) \longrightarrow 1, \tag{4.3.2.4}$$

where pr denotes the reduction modulo $\mathfrak{p}_F$:

$$\mathrm{pr}((a_{i,j})_{1\leq i,j\leq N}) := (\overline{a_{i,j}})_{1\leq i,j\leq N}, \tag{4.3.2.5}$$

where $a \mapsto \overline{a}$ is the canonical projection from $\mathfrak{o}_F$ to the finite field $k_F = \mathfrak{o}_F/\mathfrak{p}_F$.

Definition 4.3.3. A conjugate under $\mathrm{GL}_N(F)$ of the inverse image of a Borel subgroup is an *Iwahori subgroup*. A conjugate under $\mathrm{GL}_N(F)$ of the inverse image in K_0 of a parabolic subgroup of $\mathrm{GL}_N(k_F)$ is a *parahoric subgroup* of $\mathrm{GL}_N(F)$.

Thus, an Iwahori subgroup is conjugate to

$$\begin{pmatrix} \mathfrak{o}_F^\times & & \mathfrak{o}_F \\ & \ddots & \\ \mathfrak{p}_F & & \mathfrak{o}_F^\times \end{pmatrix} \tag{4.3.2.6}$$

(see also Example 4.3.15) and, more generally, a parahoric subgroup is conjugate to

$$\begin{pmatrix} \mathrm{GL}_{N_1}(\mathfrak{o}_F) & & \mathfrak{o}_F \\ & \ddots & \\ \mathfrak{p}_F & & \mathrm{GL}_{N_m}(\mathfrak{o}_F) \end{pmatrix}, \tag{4.3.2.7}$$

where $N_1 + N_2 + \cdots + N_m = N$. In particular, the group K_0 is a parahoric subgroup of $\mathrm{GL}_N(F)$. A parahoric subgroup K^P has a filtration by subgroups with similar properties as Properties 4.3.2, and we have an exact sequence

$$1 \longrightarrow K^P_{0+} \longrightarrow K^P \xrightarrow{\mathrm{pr}} \prod_{i=1}^{m} \mathrm{GL}_{N_i}(k_F) \longrightarrow 1, \tag{4.3.2.8}$$

where K^P_{0+} is called the *pro p unipotent radical* of K_P.

4.3.3 Structure of p-Adic Reductive Groups

Definition 4.3.4. The group G of the F-rational points of a connected reductive algebraic group $\mathbf{G}$ defined over F is called a *p-adic reductive group.*

Example 4.3.5. The group $\mathrm{SL}_N(F)$ is defined to be the subgroup of $\mathrm{GL}_N(F)$ formed by the elements of determinant equal to 1. The groups $\mathrm{GL}_N(F)$ and $\mathrm{SL}_N(F)$ are examples of reductive p-adic groups.

An algebraic F-group $\mathbf{T}$ is called a *torus* if $\mathbf{T}$ is isomorphic to $\mathbb{G}_{\mathrm{m}}^r$ over F_{s}. The integer r (which is also the dimension of $\mathbf{T}$) is called the *absolute rank* of $\mathbf{T}$. The *F-rank* of $\mathbf{T}$ is the largest integer s such that there is an embedding $\mathbb{G}_{\mathrm{m}}^s \hookrightarrow \mathbf{T}$ defined over F. If $\mathbf{T}$ is isomorphic to $\mathbb{G}_{\mathrm{m}}^r$ over F, we say $\mathbf{T}$ is a *split torus* (this happens precisely when $r = s$). On the other hand, when $s = 0$, we say that $\mathbf{T}$ is F-anisotropic.

For $\mathbf{G}$, a connected reductive algebraic group defined over F and $\mathbf{T} \subset \mathbf{G}$ a torus defined over F, we denote by $R(\mathbf{G},\mathbf{T})$ the set of roots of $\mathbf{G}$ with respect to $\mathbf{T}$. The group $\mathbf{G}$ is said to be almost simple if it has no nontrivial proper closed normal F-subgroups. A simple F-group is necessarily semisimple. If $\mathbf{G}$ remains simple over any extension F, then $\mathbf{G}$ is said to be absolutely (almost) simple.

An F-character of $\mathbf{G}$ is a morphism of algebraic groups $\mathbf{G} \to \mathbb{G}_{\mathrm{m}}$ defined over F. The set $X_F(\mathbf{G})$ of F-characters of $\mathbf{G}$ forms an abelian group under pointwise multiplication, called the group of characters of $\mathbf{G}$. We write $X(\mathbf{G})$ for the group $X_{F_{\mathrm{s}}}(\mathbf{G})$. An F-cocharacter of $\mathbf{G}$ is a morphism of algebraic F-groups $\mathbb{G}_{\mathrm{m}} \to \mathbf{G}$. If $\mathbf{G}$ is commutative, then the set $X_{*,F}(\mathbf{G})$ of F-cocharacters forms an abelian group under pointwise multiplication, called the group of cocharacters of $\mathbf{G}$. Again, we abbreviate $X_{*,F_s}(\mathbf{G})$ to $X_*(\mathbf{G})$. There is a pairing $\langle\ \ \rangle\colon X_F(\mathbf{G}) \times X_{*,F}(\mathbf{G}) \to \mathbb{Z}$ defined as follows. Given an F-character χ and an F-cocharacter λ, the composition $\chi \circ \lambda$ defines an F-endomorphism of the group $\mathbb{G}_\mathrm{m}$. There is thus an integer n such that $\chi \circ \lambda$ is the map $a \mapsto a^n$, and we write $\langle \chi, \lambda \rangle = n$.

The rank of any maximal F-split torus of $\mathbf{G}$ is called the rank of $\mathbf{G}$ over F (or the F-rank of $\mathbf{G}$). The absolute rank of $\mathbf{G}$ is the F_s-rank of $\mathbf{G}$. If the F-rank equals the absolute rank (i.e., if some maximal torus contained in $\mathbf{G}$ splits over F), then we say that $\mathbf{G}$ is F-split. On the other hand, if the F-rank of $\mathbf{G}$ is 0, we say that $\mathbf{G}$ is F-anisotropic.

Example 4.3.6. The subgroup

$$T := \left\{ \left(\begin{smallmatrix} a & 0 \\ 0 & a^{-1} \end{smallmatrix} \right) \mid a \in F^\times \right\} \tag{4.3.3.1}$$

of $\mathrm{SL}_2(F)$ is topologically isomorphic to $F^\times$. It is the group of F-rational points of a maximal split torus of SL_2. Let

$$U := \left\{ \left(\begin{smallmatrix} 1 & b \\ 0 & 1 \end{smallmatrix} \right) \mid a \in F^\times, b \in F \right\} \quad \text{and} \quad \overline{U} := \left\{ \left(\begin{smallmatrix} 1 & 0 \\ c & 1 \end{smallmatrix} \right) \mid a \in F^\times, c \in F \right\}. \tag{4.3.3.2}$$

The group $B := TU$ is a *Borel subgroup* of $\mathrm{SL}_2(F)$, it is a semidirect product of T and U, with U being the normal subgroup. The group $\overline{B} := T\overline{U}$ is called the *opposite* Borel subgroup.

Example 4.3.7. In $\mathrm{SL}_2(\mathbb{Q}_p)$, all non-$\mathbb{Q}_p$-split tori are totally anisotropic and split over a quadratic extension E of $\mathbb{Q}_p$. The field $F := \mathbb{Q}_p$ has three quadratic extensions. For any quadratic extension E/F, the quotient $E^\times/(E^\times)^2$ can be represented by $\{1, \varepsilon, \varpi_F, \varepsilon\varpi_F\}$, where ε is a fixed nonsquare in $\mathfrak{o}_F^\times$ (which is chosen to be -1 when $-1 \notin (E^\times)^2$), and $\varpi_F = p$. The extension E has the form $F(\sqrt{a})$, where a is ε, ϖ_F, or $\varepsilon\varpi_F$. The extension $F(\sqrt{a})/F$ is unramified if $a = \varepsilon$ and is ramified in the other two cases. A torus is called *unramified* if E is unramified, and *ramified* otherwise. Let $u, v \in F$. Setting

$$t_{u,v}(a,b) := \begin{pmatrix} a & ub \\ vb & a \end{pmatrix}, \quad \text{where } a, b \in \mathfrak{o}_F, \tag{4.3.3.3}$$

we define

$$T_{u,v} := \left\{ t_{u,v}(a,b) \ :\ a, b \in \mathfrak{o}_F \text{ such that } a^2 - uvb^2 = 1 \right\}. \tag{4.3.3.4}$$

We give a list of representatives of conjugacy classes of anisotropic tori in Table 4.1 There are six when $-1 \in (F^\times)^2$; otherwise, we have $T_{1,\varpi_F} \simeq T_{-1,-\varpi_F}$ and $T_{1,-\varpi_F} \simeq T_{-1,\varpi_F}$ and there are only four conjugacy classes.

Table 4.1 *Anisotropic tori in* $\mathrm{SL}_2(F)$

$T_{u,v}$	E	type of E	x
$T_{1,\varepsilon}$	$F(\sqrt{\varepsilon})$	unramified	0
$T_{\varpi_F^{-1},\varepsilon\varpi_F}$	$F(\sqrt{\varepsilon})$	unramified	1
T_{1,ϖ_F}	$F(\sqrt{\varpi_F})$	ramified	1/2
$T_{\varepsilon,\varepsilon^{-1}\varpi_F}$	$F(\sqrt{\varpi_F})$	ramified	1/2
$T_{1,\varepsilon\varpi_F}$	$F(\sqrt{\varepsilon\varpi_F})$	ramified	1/2
$T_{\varepsilon,\varpi}$	$F(\sqrt{\varepsilon\varpi_F})$	ramified	1/2

An F-subgroup $\mathbf{P}$ of $\mathbf{G}$ is called a *parabolic subgroup* if the quotient variety $\mathbf{G}/\mathbf{P}$ is complete. If $\mathbf{B}$ is a parabolic subgroup of $\mathbf{G}$ that is solvable, then $\mathbf{B}$ is called a *Borel subgroup* of $\mathbf{G}$.

Parabolic subgroups always exist (for example, $\mathbf{G}$ itself is a parabolic F-subgroup). However, Borel F-subgroups need not exist in general. When $\mathbf{G}$ is F-split, Borel F-subgroups always exist. The group $\mathbf{G}$ is said to be *quasi-split* over F if Borel F-subgroups do exist. In this case, one shows that the Borel F-subgroups are precisely the minimal parabolic subgroups and are also precisely the maximal connected solvable subgroups of $\mathbf{G}$.

Let $\mathbf{P}$ be a parabolic subgroup of $\mathbf{G}$. The set of all closed connected normal unipotent subgroups of $\mathbf{P}$ has a unique maximal element called the *unipotent radical* of $\mathbf{P}$. There exists a reductive subgroup $\mathbf{L}$ of $\mathbf{G}$, called a *Levi factor* of $\mathbf{P}$, such that $\mathbf{L}$ normalizes $\mathbf{U}$ and $\mathbf{P} = \mathbf{L} \ltimes \mathbf{U}$.

If $\mathbf{G}$ admits a Borel F-subgroup, then the parabolic F-subgroups of $\mathbf{G}$ are characterized as those F-subgroups of $\mathbf{G}$ that contain a Borel subgroup, and their groups of F-rational points are called *parabolic subgroups of* G. The unipotent radical $\mathbf{U}$ of an F-parabolic subgroup $\mathbf{P}$ of $\mathbf{G}$ is F-rational. The group U of F-rational points of $\mathbf{U}$ is called the unipotent radical of P (the group of F-rational points of $\mathbf{P}$). The group P decomposes as a semidirect product $P = L \ltimes U$, where the group L (called a Levi factor of P) is the group of F-rational points of $\mathbf{L}$. We have a canonical isomorphism $L \simeq P/U$ obtained by composing the injection $L \hookrightarrow P$ with the quotient map $P \to P/U$. Then P is a closed subgroup of G with a compact quotient G/P and we can induce representations from P to G. In this case, there is no difference between smooth induction and compact induction.

Example 4.3.8. For $G = \mathrm{GL}_N(F)$, up to conjugacy, a parabolic subgroup P is a subgroup of upper block-triangular matrices, U is formed by those matrices whose diagonal blocks are identity matrices, and L is the subgroup of block-diagonal matrices, a product of smaller $\mathrm{GL}_{N_i}(F)$.

Example 4.3.9. Let $\mathbf{G} = \mathrm{SL}_N$. The F-subgroup $\mathbf{T}$ consisting of diagonal matrices is a maximal F-split torus. As $\mathbf{T}$ is maximal among all tori ($\mathbf{T}$ is its own centralizer), we note that the group $\mathrm{SL}_N(F)$ is split. The characters

$$\chi_i \colon \begin{pmatrix} t_1 & & \\ & \ddots & \\ & & t_N \end{pmatrix} \mapsto t_i$$

form a basis of the free abelian group $X_F(\mathbf{T})$. The cocharacters

$$x \mapsto \mathrm{diag}\begin{pmatrix} 1 & & & & & & \\ & \ddots & & & & & \\ & & 1 & & & & \\ & & & x & & & \\ & & & & x^{-1} & & \\ & & & & & 1 & \\ & & & & & & \ddots \\ & & & & & & & 1 \end{pmatrix}$$

(with x in the i-th position, for $1 \leq i \leq N-1$) form a basis of $X_F^*(\mathbf{T})$.

Let $\mathbf{T}$ be a maximal F-split torus of $\mathbf{G}$. We let $\mathbf{T}$ act on the Lie algebra $\mathfrak{g}$ of $\mathbf{G}$ via the adjoint map. Since $\mathbf{T}$ consists of commuting semisimple elements, $\mathfrak{g}$ splits as a (finite) direct sum of eigenspaces for $\mathrm{Ad}(\mathbf{T})$, which we index by their weights $\alpha \in R_F(\mathbf{G},\mathbf{T}) \cup \{0\} \subset X_F(\mathbf{S})$:

$$\mathfrak{g} = \mathfrak{g}_0 \oplus \bigoplus_{\alpha} \mathfrak{g}_\alpha, \tag{4.3.3.5}$$

where $\mathfrak{g}_\alpha \neq 0$ and $\mathrm{Ad}(t)(X) = \alpha(t)X$ for any $t \in \mathbf{T}$, $X \in \mathfrak{g}_\alpha$ and $\alpha \in R_F(\mathbf{G},\mathbf{T})$.

One shows that $\mathfrak{g}_0$ is the Lie algebra of the centralizer $Z_{\mathbf{G}}(\mathbf{T})$ of $\mathbf{T}$ in $\mathbf{G}$. The set $R_F(\mathbf{G},\mathbf{T})$ of nonzero weights of $\mathrm{Ad}(\mathbf{T})$ is called the set of F-roots of $\mathbf{G}$ with respect to $\mathbf{T}$. It is an irreducible root system in the vector space $X_F(\mathbf{T}) \otimes_{\mathbb{Z}} \mathbb{R}$. Moreover precisely, there exists a set $R_F^\vee(\mathbf{T}) \subset X_{F,*}(\mathbf{T})$ of F-cocharacters called coroots, and a bijection $R_F(\mathbf{G},\mathbf{T}) \to R_F^\vee(\mathbf{G},\mathbf{T})$ with the property that

$$(X_F(\mathbf{T}), R_F(\mathbf{G},\mathbf{T}), X_{F,*}(\mathbf{T}), R_F^\vee(\mathbf{G},\mathbf{T}))$$

is a root datum. (In particular, one has $\langle \alpha, \alpha^\vee \rangle = 2$.) This root system (resp. root datum) is reduced if $\mathbf{G}$ is F-split.

For each $\alpha \in R_F(\mathbf{G},\mathbf{T})$, there exists a unique unipotent F-subgroup $\mathbf{U}_\alpha$ of $\mathbf{G}$, called the *root group* associated with α, whose Lie algebra is $\mathfrak{g}_\alpha \oplus \mathfrak{g}_{2\alpha}$ ($= \mathfrak{g}_\alpha$ if 2α is not a root), see [42]. The root groups are normalized by $\mathbf{T}$ and their commutators are subject to the rule $[\mathbf{U}_\alpha, \mathbf{U}_\beta] \subset \mathbf{U}_{\alpha+\beta}$ (by convention, if $\chi \in X_F(\mathbf{T})$ is not a root, we set $\mathbf{U}_\chi := \{1\}$).

Example 4.3.10. The Lie algebra of $\mathrm{SL}_N(F)$ can be identified with the Lie algebra of matrices with trace zero. If we denote by $e_{i,j}$ the (i,j)th elementary (nilpotent) matrix (which belongs to ${}_N(F)$ as soon as $i \neq j$), we see that $\mathrm{Ad}(a)(e_{i,j}) = \chi_i(a)/\chi_j(a)^{-1}$. Therefore, the set of roots of $\mathrm{SL}_N(F)$ is precisely

$$R_F(\mathrm{SL}_N, \mathbf{T}) = \{\chi_i \chi_j^{-1} \ : i \neq j\}.$$

One can show that the set of coroots is

$$R^\vee(\mathrm{SL}_N, \mathbf{T}) = \{(\lambda_i \lambda_{i+1} \ldots \lambda_j)^{\pm 1} \ : i < j\},$$

where

$$\lambda_i(x) = \begin{pmatrix} 1 & & & & & \\ & \ddots & & & & \\ & & 1 & & & \\ & & & x & & \\ & & & & 1 & \\ & & & & & \ddots & \\ & & & & & & 1 \end{pmatrix},$$

with x in the i-th position, and $R_F(\mathrm{SL}_N, \mathbf{T})$ is the root system of type A_{n-1} (see Example 4.3.12). For $i \neq j$, let $u_{i,j}(a)$ be the elementary unipotent matrix with a as its (i,j)th entry. Then $u_{i,j}$ defines an isomorphism $a \mapsto u_{i,j}(a)$ from $\mathbb{G}_a$ onto $\mathbf{U}_{i,j} := \mathbf{U}_{\alpha_{i,j}}$, the root group associated to the root $\alpha_{i,j} := \chi_i \chi_j^{-1}$. The parabolic subgroups containing $\mathbf{T}$ can then be obtained as the groups generated by $\mathbf{T}$ and some appropriate subset of the root groups $\mathbf{U}_{i,j}$ with $i \neq j$.

The normalizer $\mathrm{N}_{\mathbf{G}}(\mathbf{T})$ of $\mathbf{T}$ in $\mathbf{G}$ acts on $\mathfrak{g}$ via the adjoint action, and (equivalently) on the set of root groups via conjugation, with the centralizer $\mathrm{Z}_{\mathbf{G}}(\mathbf{T})$ of $\mathbf{T}$ in $\mathbf{G}$ acting trivially. The induced map $\mathrm{N}_{\mathbf{G}}(\mathbf{T})(F)/\mathrm{Z}_{\mathbf{G}}(\mathbf{T})(T) \to W$ is an isomorphism onto the Weyl group W of the root system $R_F(\mathbf{G}, \mathbf{T})$. In particular, we can find representatives in $\mathrm{N}_{\mathbf{G}}(\mathbf{T})(F)$ for the action of W.

There is a W-equivariant bijection between bases of the root system $R_F(\mathbf{G}, \mathbf{T})$ and minimal parabolic F-subgroups of $\mathbf{G}$ containing $\mathbf{T}$. Indeed, given a basis $\Delta := \{\alpha_1, \ldots, \alpha_r\}$ of $R_F(\mathbf{G}, \mathbf{T})$, the subgroup of $\mathbf{G}$ generated by $\mathrm{Z}_{\mathbf{G}}(\mathbf{T})$ and the root groups $\mathbf{U}_\alpha$ for $\alpha \in \Delta$ is a minimal F-parabolic of $\mathbf{G}$ containing $\mathbf{T}$. Conversely, if $\mathbf{P}$ is a minimal F-parabolic containing $\mathbf{T}$, its unipotent radical is normalized by $\mathbf{T}$, hence the Lie algebra of the latter decomposes as a sum $\bigoplus_{\alpha \in R_F(\mathbf{G},\mathbf{T})^+} \mathfrak{g}_\alpha$ of root spaces. The index set $R_F(\mathbf{G}, \mathbf{T})^+$ defines a set of positive roots in $R_F(\mathbf{G}, \mathbf{T})$ and W acts simply transitively on the set of minimal parabolic subgroups containing $\mathbf{T}$.

4.3.4 Finite and Affine Weyl Groups

The notion of *Coxeter groups* first appeared in Tits' 1961 mimeographed notes, *Groupes et géométries de Coxeter* (they were reproduced in [36]).

Definition 4.3.11. A *Coxeter group* W is a group generated by a set S of elements of order 2, which has a presentation

$$W = \langle S \ : \ (ss')^{m(s,s')} = 1 \text{ for all } s, s' \in S\rangle,$$

where $m(s,s') \in \mathbb{Z}_{\geq 1}$ is the order of ss' in W.

The equalities $s^2 = 1$ (for $s \in S$) are called the *quadratic relations*, while the equalities $(ss')^{m(s,s')} = 1$, or equivalently

$$\underbrace{ss'ss'\cdots}_{m(s,\,s')\text{ terms}} = \underbrace{s'ss's\cdots}_{m(s,\,s')\text{ terms}} \qquad (4.3.4.1)$$

are called the *braid relations*.

Example 4.3.12. Examples of finite Coxeter groups:

1. $W = S_n$, with $S = \{(12), (23), \ldots, (n-1n)\}$, symmetric group, type A_{n-1};
2. $W = S_n \ltimes \{\pm 1\}^n$, with $S = \{(12), (23), \ldots, (n-1n), (\mathrm{id}, (1, \ldots, 1, -1))\}$, hyperoctahedral group, type B_n or C_n.

In a Coxeter group, the *length* $\ell(w)$ of $w \in W$ is defined to be the minimum number of generators in an expression of w.

Let $\mathscr{R} := (X, R, Y, R^\vee)$ be a root datum, that is,

- X and Y are lattices of finite rank, with a perfect pairing $\langle\,,\,\rangle \colon X \times Y \to \mathbb{Z}$,
- R is a root system in X,
- $R^\vee \subset Y$ is the dual root system of R, and a bijection $R \to R^\vee$, $\alpha \mapsto \alpha^\vee$ with $\langle \alpha, \alpha^\vee\rangle = 2$ is given,
- for every $\alpha \in R$, the reflection $s_\alpha \colon X \to X$ defined by $s_\alpha(x) := x - \langle x, \alpha^\vee\rangle\alpha$ stabilizes R,
- for every $\alpha \in R$, the reflection $s_{\alpha^\vee} \colon Y \to Y$ defined by $s_{\alpha^\vee}(y) := y - \langle \alpha, y\rangle\alpha^\vee$ stabilizes $R^\vee$.

We denote by W the group generated by the s_α for $\alpha \in R$. It is a finite Weyl group. We write $S := \{s_\alpha \ : \ \alpha \in \Delta\}$. Then (W, S) is a finite Coxeter system.

Given an ordered vector space structure on $X \otimes_{\mathbb{Z}} \mathbb{R}$ such that any root is positive or negative, we denote by R^+ (resp. R^-) the set of positive (resp. ngeative) roots. Positive roots that are indecomposable into a sum of other positive roots are called *simple roots*. The set Δ of simple roots is called the *basis* of R relative to the chosen order.

The length $\ell(w)$ of $w \in W$ is equal to the number of roots in R^+ that are sent to R^- under the action of W (see [28, VI, Section 1.6]).

We set $\widetilde{W} := W \ltimes X$. The group $\widetilde{W}$ is called the *extended affine Weyl group* of $\mathscr{R}$. For $\alpha \in R$ and $n \in \mathbb{Z}$, we define the hyperplane

$$H_{\alpha,n} := \{x \in X \otimes_{\mathbb{Z}} \mathbb{R} \,:\, \alpha^\vee(x) = n\}. \qquad (4.3.4.2)$$

Let $W_{\rm aff}$ be the subgroup of $\widetilde{W}$ generated by the (affine) reflections in the hyperplanes $H_{\alpha,n}$, for $\alpha \in R$ and $n \in \mathbb{Z}$. This is an *affine Weyl group*, in particular, it is a Coxeter group, and $X \otimes_{\mathbb{Z}} \mathbb{R}$ with this hyperplane arrangement is the *Coxeter complex* of $W_{\rm aff}$ (see [75, Chap. 2]).

Let A_0 denote the unique alcove contained in the positive Weyl chamber (with respect to Δ), such that $0 \in \overline{A_0}$. The reflections in those walls of A_0 that contain 0 constitute precisely S. We denote by $S_{\rm aff}$ the set of affine reflections with respect to all walls of A_0. The length $\ell(w)$ of $w \in W_{\rm aff}$ is equal to the number of hyperplanes $H_{\alpha,n}$ that separate $w(A_0)$ from A_0 (see [50, Section 4.4]).

Let $R^\vee_{\max}$ be the set of maximal elements of $R^\vee$, with respect to the base $\Delta^\vee := \{\alpha^\vee \,:\, \alpha \in \Delta\}$. It contains one element for every irreducible component of $R^\vee$. For $\alpha^\vee \in R^\vee_{\max}$ we define

$$s'_\alpha \colon X \to X,\ s'_\alpha(x) := x + \alpha - \langle x, \alpha^\vee\rangle\alpha.$$

This is the reflection of $X \otimes_{\mathbb{Z}} \mathbb{R}$ in the hyperplane $H_{\alpha,1}$. Then $S_{\rm aff} = S \cup \{s'_\alpha \,:\, \alpha^\vee \in R^\vee_{\max}\}$.

4.3.5 Affine Hecke Algebras

We keep the notation of Section 4.3.4. In the group algebra $\mathbb{C}[W]$, the quadratic relations are equivalent to

$$(s+1)(s-1) = 0 \quad \text{for any } s \in S. \qquad (4.3.5.1)$$

We choose, for every $s \in S$, a complex number q_s, such that

$$q_s = q_{s'} \quad \text{if } s \text{ and } s' \text{ are conjugate in } W. \qquad (4.3.5.2)$$

Let $\mathbf{q} \colon S \to \mathbb{C}$ be the function $s \mapsto q_s$. We define a new $\mathbb{C}$-algebra $\mathscr{H}(W,q)$, which has a vector space basis $\{T_w \,:\, w \in W\}$. Here T_1 is the unit element and the T_W satisfy the following quadratic relations and braid relations

$$(T_s+1)(T_s-q_s) = 0 \quad \text{and} \quad \underbrace{T_sT_{s'}T_s\cdots}_{m(s,s')\text{ terms}} = \underbrace{T_{s'}T_sT_{s'}\cdots}_{m(s,s')\text{ terms}} \quad \text{for any } s, s' \in S. \qquad (4.3.5.3)$$

Fix $q \in \mathbb{R}_{>1}$ and let $\lambda, \lambda^* \colon R \to \mathbb{C}$ be functions such that

- if s_α and s_β are conjugate in W, then $\lambda(\alpha) = \lambda(\beta)$ and $\lambda^*(\alpha) = \lambda^*(\beta)$,
- if $\alpha^\vee \notin 2Y$, then $\lambda^*(\alpha) = \lambda(\alpha)$.

We note that $\alpha^\vee \in 2Y$ is only possible for short roots α in a type B component of the root system R.

For $\alpha \in R$, we write

$$q_{s_\alpha} := q^{\lambda(\alpha)} \quad \text{and (if } \alpha^\vee \in R^\vee_{\max}) \quad q_{s'_\alpha} := q^{\lambda^*(\alpha)}. \tag{4.3.5.4}$$

We denote by ℓ the usual length function on W. Let $\mathcal{H}(W,\mathbf{q})$ denote the Iwahori–Hecke algebra of W. It has a basis $\{T_w : w \in W\}$ such that

$$\begin{aligned} T_w T_v &= T_{wv} \quad \text{if } \ell(w) + \ell(v) = \ell(wv), \\ (T_{s_\alpha} + 1)(T_{s_\alpha} - q_{s_\alpha}) &= 0 \qquad \text{if } \alpha \in \Delta. \end{aligned}$$

Let $\{\theta_x : x \in X\}$ denote the standard basis of $\mathbb{C}[X]$. Then the *affine Hecke algebra* $\mathcal{H}(\mathcal{R}, \lambda, \lambda^*, q)$ is the vector space $\mathbb{C}[X] \otimes_{\mathbb{C}} \mathcal{H}(W,\mathbf{q})$ such that $\mathbb{C}[X]$ and $\mathcal{H}(W,\mathbf{q})$ are embedded as subalgebras, and for $\alpha \in \Delta$ and $x \in X$:

$$\theta_x T_{s_\alpha} - T_{s_\alpha}\theta_{s_\alpha(x)} = \left((q^{\lambda(\alpha)} - 1) + \theta_{-\alpha}\left(q^{\frac{\lambda(\alpha)+\lambda^*(\alpha)}{2}} - q^{\frac{\lambda(\alpha)-\lambda^*(\alpha)}{2}}\right)\right)\frac{\theta_x - \theta_{s_\alpha(x)}}{\theta_0 - \theta_{-2\alpha}}.$$

Remark 4.3.13. When $\alpha^\vee \notin 2Y$, the cross relation simplifies to

$$\theta_x T_{s_\alpha} - T_{s_\alpha}\theta_{s_\alpha(x)} = (q^{\lambda(\alpha)} - 1)\frac{\theta_x - \theta_{s_\alpha(x)}}{\theta_0 - \theta_{-\alpha}}.$$

4.3.6 Graded Affine Hecke Algebras

Graded affine Hecke algebras were defined by Lusztig [61] in his study of unipotent representations of reductive p-adic groups. In [20], a Dirac operator for graded Hecke algebras $\mathbb{H}$ was defined and, by analogy with the setting of Dirac theory for $(\mathfrak{g}, K)$-modules of real reductive groups, the notion of Dirac cohomology was introduced. The Dirac cohomology and the Dirac index in the Hecke algebra setting were further studied in [37] and [39]. The Dirac cohomology spaces are representations for a certain double cover ("pin cover") $\widetilde{W}$ of the Weyl group W. The irreducible representations of $\widetilde{W}$ had been classified case by case in the work of Schur, Morris, Read, and others. Recently, it was noticed in [38] that there is a close relation between the representation theory of $\widetilde{W}$ and the geometry of the nilpotent cone in semisimple Lie algebras $\mathfrak{g}$.

Fix a root system $R = (V_0, R, V_0^\vee, R^\vee)$ over the real numbers. In particular, the set of roots R is contained in $V_0\backslash\{0\}$ and spans V_0, while the set of coroots $R^\vee \subset V_0^\vee\backslash\{0\}$ and spans $V_0^\vee$, there is a perfect bilinear pairing

$$(\,,\,)\colon V_0 \times V_0^\vee \to \mathbb{R},$$

and a bijection between R and $R^\vee$ denoted $\alpha \mapsto \alpha^\vee$ such that $(\alpha, \alpha^\vee) = 2$. We write $V := V_0 \otimes_{\mathbb{R}} \mathbb{C}$ and $V^\vee := V_0 \otimes_{\mathbb{R}} \mathbb{C}$, they are the complex spans of R and $R^\vee$.

We suppose that the root system R is reduced, meaning that $\alpha \in R$ implies that $2\alpha \notin R$.

For $\alpha \in R$, we define reflections $s_\alpha \colon V_0 \to V_0$ and $s_\alpha^\vee \colon V_0^\vee \to V_0^\vee$ by setting

$$s_\alpha(v) := v - (v,\alpha^\vee)\alpha \quad \text{and} \quad s_\alpha^\vee(v') := v' - (\alpha,v')\alpha^\vee \quad \text{for } v \in V_0,\, v' \in V_0^\vee.$$

Let W be the subgroup of $\mathrm{GL}(V_0)$ generated by $\{s_\alpha \,:\, \alpha \in R\}$.

We fix a W-invariant $c \colon R \to \mathbb{R}_{>}0$, and set $c_\alpha := c(\alpha)$.

Definition 4.3.14. [61] The *graded affine Hecke algebra* $\mathbb{H} = \mathbb{H}(R,c)$ attached to the root system R and with parameter function c is the complex associative algebra with unit generated by the symbols $\{t_w \,:\, w \in W\}$ and $\{t_f \,:\, f \in S(V^\vee)\}$, subject to the following relations:

(1) The linear map from the group algebra $C[W] = \bigoplus_{w \in W} C_w$ to $\mathbb{H}$ taking w to t_w is an injective map of algebras.
(2) The linear map from $S(V^\vee)$ to $\mathbb{H}$ taking an element f to t_f is an injective map of algebras.
(3) For all $\alpha \in \Delta$ and $\omega \in V^\vee$, we have

$$\omega t_{s_\alpha} - t_{s_\alpha} s_\alpha(\omega) = c_\alpha(\alpha,\omega),$$

where $s_\alpha(\omega)$ is the element of $V^\vee$ obtained by s_α acting on ω.

The center $\mathrm{Z}(\mathbb{H})$ of $\mathbb{H}$ is $S(V^\vee)$ (see [61, Proposition 4.5]).
For every $\omega \in V^\vee$, we set

$$\widetilde{\omega} := \omega - \frac{1}{2}\sum_{\beta>0} c_\beta(\beta,\omega)t_{s_\beta} \in \mathbb{H}. \tag{4.3.6.1}$$

Using the bilinear pairing $(\cdot,\cdot)$, we define a dual inner product on V_0 as follows. Let be $(\omega_i)_{i=1}^n$ and $(\omega^j)_{j=1}^n$ be the R-bases of $V_0^\vee$ which are in duality; that is, such that $\langle\omega_i,\omega^j\rangle = \delta_{i,j}$, the Kronecker delta. Then for $v_1,v_2 \in V_0$, we set

$$\langle v_1,v_2\rangle := \sum_{i=1}^{n}(v_1,\omega_i)(v_2,\omega^i). \tag{4.3.6.2}$$

Then (4.3.6.2) defines a W-invariant inner product on V_0. It does not depend on the choice of bases $(\omega_i)_{i=1}^n$ and $(\omega^j)_{j=1}^n$. If v is a vector in V or in $V^\vee$, we set $|v| := \langle v,v\rangle^{1/2}$.

4.3.7 Spherical Building and Bruhat–Tits Building

The notion of *building* was first introduced by Jacques Tits (see [88], [1], [75]) as a means of understanding algebraic groups over an arbitrary field $\mathbb{F}$. The general idea is to construct a space upon which the group acts in a nice manner and to use information about this space and the action to learn about the group itself. Tits described how, given a simple algebraic group $\mathbf{G}$, one can construct an associated simplicial complex $\mathscr{I}(\mathbf{G})$ with a canonical $\mathbf{G}$-action. Furthermore, the construction is natural in the sense that any group homomorphism $\mathbf{G} \to \mathbf{G}'$ induces a morphism of simplicial complexes $\mathscr{I}(\mathbf{G}) \to \mathscr{I}(\mathbf{G}')$.

A *simplicial complex* is the "gluing together" of constituent elements called *simplices* in a manner that satisfies certain axioms.

- A 0-simplex (or *vertex*) is a point.
- A 1-simplex is a line connecting two points.
- A 2-simplex is a triangle filling in the area between three lines.
- A 3-simplex is a tetrahedron which is the volume within four triangles (two cells).
- An *n-simplex* is the connecting space for $(n + 1)$ many $(n - 1)$-simplices.

The main axiom concerning gluing is that simplicies may only be glued along sub-simplicies.

The simplicial complex $\mathscr{I}(\mathbf{G})$ constructed by Tits is called the *spherical building* of $\mathbf{G}$. When $\mathbf{G}$ is semisimple, it is defined as follows:

- The vertices are the maximal parabolic subgroups of $\mathbf{G}$;
- The vertices $\mathbf{P}_0, \mathbf{P}_1, \ldots, \mathbf{P}_n$ form an n-simplex if and only if $\mathbf{P}_0 \cap \mathbf{P}_1 \cap \cdots \mathbf{P}_n$ is a parabolic subgroup of $\mathbf{G}$.

A simplex $\mathbf{P}'$ is called a *facet* of a simplex $\mathbf{P}$ if $\mathbf{P}' \subset \mathbf{P}$. The maximal facets are called *chambers*. The group $\mathbf{G}$ acts on $\mathscr{I}(\mathbf{G})$ by conjugation, and this action is transitive on the chambers. In particular, all chambers have the same dimension d. The *rank* of $\mathscr{I}(\mathbf{G})$ is the integer $d + 1$; it equals the $\mathbb{F}$-rank of $\mathbf{G}$.

The simplicial complex $\mathscr{I}(\mathbf{G})$ is a building, in the sense that there exists a collection of subcomplexes called apartments such that

1. each apartment $\mathscr{A}$ is a Coxeter complex,
2. $\mathscr{I}(\mathbf{G})$ is the union of all its apartments;
3. any two simplices $\mathbf{P}_1$ and $\mathbf{P}_2$ of $\mathscr{I}(\mathbf{G})$ are contained in a common apartment;
4. if two simplices $\mathbf{P}_1$ and $\mathbf{P}_2$ are contained in two apartments $\mathbf{A}_1$ and $\mathbf{A}_2$, then there exists an isomorphism $\mathscr{A}_1 \to \mathscr{A}_2$ fixing $\mathbf{P}_1$ and $\mathbf{P}_2$ pointwise.

The spherical building $\mathscr{I}(\mathbf{G})$ has a canonical system of apartments, indexed by the set of maximal $\mathbb{F}$-split tori of $\mathbf{G}$: To each $\mathbb{F}$-split torus $\mathbf{T}$ of $\mathbf{G}$, one associates the subcomplex $\mathscr{A}(\mathbf{G},\mathbf{T})$ consisting of all the $\mathbb{F}$-parabolic subgroups containing $\mathbf{T}$. This apartment $\mathscr{A}(\mathbf{G},\mathbf{T})$ is a Coxeter complex of the same type as the root system $R(\mathbf{G},\mathbf{T})$. In particular, the Coxeter group associated with $\mathscr{A}(\mathbf{G},\mathbf{T})$ is precisely the Weyl group W of $R(\mathbf{G},\mathbf{T})$. The normalizer $\mathrm{N}_{\mathbf{G}}(\mathbf{T})$ of $\mathbf{T}$ stabilizes the apartment $\mathscr{A}(\mathbf{G},\mathbf{T})$, the centralizer $\mathrm{Z}_{\mathbf{G}}(\mathbf{T})$ of $\mathbf{T}$ fixes $\mathscr{A}(\mathbf{G},\mathbf{T})$ pointwise, and the quotient $\mathrm{N}_{\mathbf{G}}(\mathbf{T})(\mathbb{F})/\mathrm{Z}_{\mathbf{G}}(\mathbf{T})(\mathbb{F})$ acts on $\mathscr{A}(\mathbf{G},\mathbf{T})$ through W.

If $\mathbf{G}$ is a connected reductive algebraic group over F (a non-archimedean local field), another important tool in understanding the structure of G and its representations is the (enlarged) *Bruhat–Tits building* $\mathscr{B}(\mathbf{G},F)$ of G. It is a product of a polysimplicial complex with a finite-dimensional affine space on which the group G acts (see [29, Section 7.4]).

One motivation to introduce Bruhat–Tits buildings is because they provide a very useful non-archimedean replacement of Riemannian symmetric spaces. The Riemannian symmetric space associated with a Lie group $\mathscr{G}$ is the quotient space $\mathscr{G}/K$, where K is a maximal compact subgroup of $\mathscr{G}$.

The Bruhat–Tits building $\mathscr{B}(\mathbf{G},F)$ is a set equipped with the following structures:

- it is a complete metric space, with an affine structure;
- it is the product of a polysimplicial complex and a real vector space;
- it has a collection of distinguished subsets, known as apartments, which are indexed by the maximal split tori of $\mathbf{G}$;
- the group G acts isometrically on $\mathscr{B}(\mathbf{G},F)$ as simplicial automorphisms.

The affine Weyl group $\widetilde{W}$ of $\mathbf{G}(F)$ is the Coxeter group of any of the apartments of $\mathscr{B}(\mathbf{G},F)$. It is the semidirect product of the (spherical) Weyl group W of $\mathbf{G}$ by a free abelian group of rank equal to the rank of $\mathbf{G}$.

Moreover, for any $x \in \mathscr{B}(\mathbf{G},F)$, the stabilizer $\mathrm{Stab}_G(x)$ of x in G is a compact open subgroup of G. Stabilizers in G of chambers in $\mathscr{B}(\mathbf{G},F)$ are called *Iwahori subgroups*; they are exactly the normalizers of the maximal pro-p subgroups of G.

Example 4.3.15. Let $\mathbf{B}$ be a Borel subgroup of $\mathbf{G}$, and let denote by $\overline{g}$ the reduction of $g \in \mathbf{G}(\mathfrak{o}_F)$. The subgroup

$$I := \{g \in \mathbf{G}(\mathfrak{o}_F) \ : \ \overline{g} \in \mathbf{B}(k_F)\} \tag{4.3.7.1}$$

is an Iwahori subgroup of G. When $\mathbf{G} = \mathrm{GL}_2$, we have

$$I = \left\{ \begin{pmatrix} a & b \\ c & d \end{pmatrix} \ : \ a, d \in \mathfrak{o}_F^\times, b \in \mathfrak{o}_F, c \in \mathfrak{p}_F \right\}. \tag{4.3.7.2}$$

The group G acts transitively on the set of chambers, and equivalently, all Iwahori subgroups are conjugate under G. Stabilizers in G of simplices in $\mathscr{B}(\mathbf{G}, F)$ are called parahoric subgroups; they are precisely the compact open subgroups of G containing an Iwahori subgroup. This correspondence between simplices and parahoric subgroups is bijective and reverses inclusions. The type of a parahoric subgroup is defined to be the type of the associated simplex. The stabilizers in G of vertices in $\mathscr{B}(\mathbf{G}, F)$ are the maximal compact subgroups of G.

A vertex $x \in \mathscr{B}(\mathbf{G}, F)$ is called *special* if, with respect to any apartment $\mathscr{A}$ containing x, the affine Weyl group $\widetilde{W}$ is the semidirect product of the stabilizer $\widetilde{W}_x$ of x by the translation subgroup. If so, then $\widetilde{W}_x$ is isomorphic to the (spherical) Weyl group W. If G splits over F_{nr} and x, seen as a vertex of $\mathscr{B}(\mathbf{G}, F_{\mathrm{nr}})$ is special, then it is called *hyperspecial*. Hyperspecial vertices are special. If x is a special (resp. hyperspecial) vertex, then the parahoric subgroup $G_{x,0}$ is called special (resp. hyperspecial).

Given a maximal F-split torus $\mathbf{T}$ of $\mathbf{G}$, the maximal compact subgroup Z_{c} of $\mathrm{Z}_{\mathbf{G}}(\mathbf{T})(F)$ is the fixator of an apartment $\mathscr{A}$ of $\mathscr{B}(\mathbf{G}, F)$. The group $\mathrm{N}_{\mathbf{G}}(\mathbf{T})(F)$ stabilizes $\mathscr{A}$ and the quotient $\mathrm{N}_{\mathbf{G}}(\mathbf{T})(F)/\mathrm{Z}_{\mathrm{c}}$ acts on $\mathscr{A}$ through the affine Weyl group $\widetilde{W}$. With this identification, $\mathrm{Z}_{\mathbf{G}}(\mathbf{T})(F)/\mathrm{Z}_{\mathrm{c}}$ corresponds to the translation subgroup of $\widetilde{W}$.

The Bruhat–Tits building of G collects and parametrizes a vast amount of data concerning the group G, notably that of its compact open subgroups. It is used to define filtrations both of compact open subgroups of G, of the Lie algebra of G and of its dual, all of which are crucial in the study of the representation theory of p-adic groups.

Example 4.3.16. The Bruhat–Tits building of $\mathrm{SL}_N(F)$ is simplicial and can be realized as follows (see [1, Chapter 6.9]). Let V be an N-dimensional vector space over F. An $\mathfrak{o}_F$-lattice is a free $\mathfrak{o}_F$-submodule of V of rank N. Let Latt denote the set of lattices. We say that two lattices $\mathscr{L}$ and $\mathscr{L}'$ are equivalent (denoted as $\mathscr{L} \simeq \mathscr{L}'$) if they are equivalent up to homothety, that is, if $\mathscr{L} = a\mathscr{L}'$ for some $a \in F^{\times}$. We denote by $[\mathscr{L}]$ the equivalence class of a lattice $\mathscr{L}$.

In order to realize $\mathscr{B}_N := \mathscr{B}(\mathrm{SL}_N, F)$ as a simplicial complex, we take the set Latt/ $\sim$ of equivalence classes of lattices as the set of vertices. The 1-simplexes (*edges*) of $\mathscr{B}_N$ are the lines joining $[\mathscr{L}_0]$, $[\mathscr{L}_0]$ where we have $\varpi_F \mathscr{L}_0 \subset \mathscr{L}_1 \subset \mathscr{L}_0$, for some representatives $\mathscr{L}_0$, $\mathscr{L}_1$ of the equivalence classes.

We take for n-simplices every possible complete subgraph (in the already defined 1-skeleton) on $n + 1$ many vertices. Equivalently, we can describe for instance the 2-simplices as all collections $[\mathscr{L}_0]$, $[\mathscr{L}_1]$, $[\mathscr{L}_2]$ with $\varpi_F \mathscr{L}_0 \subset$

$\mathscr{L}_2 \subset \mathscr{L}_1 \subset \mathscr{L}_0$. This generalizes for n-simplices: A collection of $(n+1)$ vertices, $\Lambda_0, \Lambda_1, \ldots, \Lambda_n$ forms an n-simplex in $\mathscr{B}_N$ if there exists lattice representatives $[\mathscr{L}_m] := \Lambda_m$ for $0 \leq m \leq n$ such that

$$\varpi_F \mathscr{L}_0 \subset \mathscr{L}_n \subset \mathscr{L}_{n-1} \subset \cdots \subset \mathscr{L}_1 \subset \mathscr{L}_0.$$

If $\Lambda_0, \Lambda_1, \ldots, \Lambda_n$ are the vertices of an n-simplex in $\mathscr{B}_N$, we denote by $\Lambda_0\Lambda_1 \cdots \Lambda_n$ the n-simplex they span. Then $\mathscr{B}_N$ is the abstract simplicial complex with the following simplices

$$S_0(\mathscr{B}_N) := \{\Lambda_m \ : \ \Lambda_m \in \mathrm{Latt}/\sim\},$$

and $S_m(\mathscr{B}_N)$ for $1 \leq m \leq N$, where $S_m(\mathscr{B}_N)$ is the set of m-simplices $\Lambda_0\Lambda_1 \cdots \Lambda_m$ such that, for each $i \in \{0, 1, \ldots, m\}$ there exists $\mathscr{L}_i \in \mathrm{Latt}$ with $[L_i] = \Lambda_i$ and $\varpi_F \mathscr{L}_0 \subset \mathscr{L}_m \subset \mathscr{L}_{m-1} \subset \cdots \subset \mathscr{L}_1 \subset \mathscr{L}_0$.

For each parahoric subgroup $G_{x,0}$ of G, there exists a smooth affine group scheme $\mathfrak{G}$ defined over $\mathfrak{o}_F$, whose group of $\mathfrak{o}_f$-points $\mathfrak{G}(\mathfrak{o}_F)$ is isomorphic to G_x, whose generic fiber $\mathfrak{G} \times_{\mathfrak{o}_F} F$ is isomorphic to G and whose special fiber $\mathbb{G}_x := \mathfrak{G} \times_{\mathfrak{o}_F} k_F$ is called the reductive quotient group of G_x. The reduction map $G_{x,0} \to \mathbb{G}_x$ is surjective.

Example 4.3.17. For $\mathbf{G} = \mathrm{SL}_2$, the apartments have dimension one. Since the geometric realization of $\mathscr{B}(\mathbf{G}, F)$ is contractible, it is a tree. Moreover, the number k of edges at a vertex x of $\mathscr{B}(\mathbf{G}, F)$ is the number of Iwahori subgroups of G contained in G_x, which equals the number of Borel subgroups of $\mathbb{G}_x(k_F) \simeq \mathrm{SL}_2(k_F)$. Since Borel subgroups are self-normalizing and conjugate, for any Borel subgroup $\mathbb{B}_x$ of $\mathbb{G}_x$, we have $k = [\mathrm{SL}_2(k_F) : \mathbb{B}_x(k_F)] = q+1$. Thus $\mathscr{B}(\mathbf{G}, F)$ is a regular tree of valence $q+1$.

4.3.7.1 Moy–Prasad Filtrations

In [67, 68], Moy and Prasad defined a filtration of the parahoric subgroup $G_{x,0}$ by subgroups

$$G_{x,0} \rhd G_{x,r_1} \rhd G_{x,r_2} \rhd G_{x,r_3} \rhd \cdots, \tag{4.3.7.3}$$

that are normal inside each other and whose intersection is trivial, where $0 < r_1 < r_2 < r_3 < \cdots$ are real numbers depending on x. For $r \in \mathbb{R}_{\geq 0}$, we write

$$G_{x,r+} := \bigcup_{t>r} G_{x,t}. \tag{4.3.7.4}$$

When $r > 0$, the quotient $G_{x,r}/G_{x,r+}$ is abelian and can be identified with a k_F-vector space. Moy and Prasad also defined filtration submodules $\mathfrak{g}_{x,r}$ of

$\mathfrak{g} := \mathrm{Lie}(\mathbf{G})(F)$ and $\mathfrak{g}^*_{x,r}$ of the linear dual $\mathfrak{g}^*$ of $\mathfrak{g}$, for $r \in \mathbb{R}$. We write $\mathfrak{g}_{x,r+} := \bigcup_{t>r} \mathfrak{g}_{x,t}$.

Note that in [91], different filtrations were introduced. As shown in [91, Section 8.1] or [51, Section 13], in the case of quasi-split groups that split over a tamely ramified extension, both notions of filtrations coincide. Related filtrations by affinoid groups, in the Berkovich analytification of a connected reductive group, have been defined in [62] (see also [71] and the references there).

Example 4.3.18. For $G = \mathrm{SL}_2(F)$, the Moy–Prasad filtration subgroups $G_{y,r}$ for $y \in \mathscr{B}(G)$ and $r \in \mathbb{R}$ are

$$G_{y,r} := \begin{pmatrix} 1 + \mathfrak{p}_F^{\lceil r \rceil} & \mathfrak{p}_F^{\lceil r-y \rceil} \\ \mathfrak{p}_F^{\lceil r+y \rceil} & 1 + \mathfrak{p}_F^{\lceil r \rceil} \end{pmatrix} \cap \mathrm{SL}_2(F), \tag{4.3.7.5}$$

where

$$\lceil r \rceil = \min \{n \in \mathbb{Z} \,:\, n \geq r\}. \tag{4.3.7.6}$$

It follows that

$$G_{[0,1/2],r/2} := G_{0,r/2} \cap G_{1/2,r/2} = \begin{pmatrix} 1 + \mathfrak{p}_F^{\lceil r/2 \rceil} & \mathfrak{p}_F^{\lceil r/2 \rceil} \\ \mathfrak{p}_F^{\lceil (r+1)/2 \rceil} & 1 + \mathfrak{p}_F^{\lceil r/2 \rceil} \end{pmatrix} \cap \mathrm{SL}_2(F). \tag{4.3.7.7}$$

The corresponding Moy–Prasad lattices are

$$\mathfrak{g}_{y,r} := \begin{pmatrix} \mathfrak{p}_F^{\lceil r \rceil} & \mathfrak{p}_F^{\lceil r-y \rceil} \\ \mathfrak{p}_F^{\lceil r+y \rceil} & \mathfrak{p}_F^{\lceil r \rceil} \end{pmatrix} \cap \mathrm{SL}_2(F). \tag{4.3.7.8}$$

Using the notation introduced in (4.3.7.6) and (4.3.1.6), we write $T_{u,v,0} := T$ and for any $r \in \widetilde{R}$:

$$T_{u,v,r} := \left\{t_{u,v}(a,b) \in T_{u,v} \,:\, a \in U^{\lceil r \rceil}(\mathfrak{o}_F) \text{ and } b \in \mathfrak{p}_F^{\lceil r-y \rceil}\right\} \tag{4.3.7.9}$$

defines the Moy–Prasad filtration of $T_{u,v}$. The Lie algebra $\mathfrak{t}$ of T is the one-dimensional subalgebra of $\mathfrak{g}$ spanned by the element $X_{u,v} := \left(\begin{smallmatrix} 0 & u \\ v & 0 \end{smallmatrix}\right)$. For any $r \in \widetilde{R}$, the corresponding filtration subring of $\mathfrak{t}$ is

$$\mathfrak{t}_{u,v,r} := \left\{\lambda X_{u,v} \,:\, \lambda \in F \text{ such that } \lambda u \in \mathfrak{p}_F^{\lceil r-y \rceil}\right\}. \tag{4.3.7.10}$$

4.4 Representations of p-Adic Reductive Groups: Basic Properties

4.4.1 Admissible Representations

Definition 4.4.1. A smooth representation (π, V) of G is *admissible* if for every compact open subgroup K of G, the space V^K is finite dimensional.

Proposition 4.4.2. If (π, V) is admissible, then its contragredient $(\tilde{\pi}, \widetilde{V})$ is admissible and we have $\tilde{\tilde{\pi}} \simeq \pi$.

Proof. Let K be a compact open subgroup of G and V_1 a K-invariant subspace of V such that $V = V^K \otimes V_1$. Let $\lambda \in \widetilde{V}^K$. Then the restriction of λ to V_1 intertwines the representation of K on V_1 and the trivial representation of K on $\mathbb{C}$. Since V_1 is a direct sum of nontrivial representations of K, if we have $\mathrm{Hom}_K(V_1, \mathbb{C}) \neq 0$, then the trivial representation of K would occur as a subrepresentation of V_1. Thus, λ restricts trivially to V_1, hence λ belongs to the finite dimensional vector space $(V^K)^*$, and hence $\widetilde{V}^K \subset (V^K)^*$, which shows that $\dim_{\mathbb{C}}(\widetilde{V}^K)$ is finite.

Let $\lambda \in (V^K)^*$. We extend it to V^* by 0 on V_1. With this extension belonging to $\widetilde{V}^K$, we obtain $\widetilde{V}^K = (V^K)^*$. This shows that $\widetilde{\widetilde{V}}^K \simeq V^K$. By letting K vary, we get $\widetilde{\widetilde{V}} \simeq V$. It follows that the natural map $V \to V$ is invertible, and thus $\tilde{\tilde{\pi}} \simeq \pi$. □

Definition 4.4.3. A *distribution* Θ on G is a linear functional on $\mathscr{C}^\infty(G)$. It is said to be G-invariant (or invariant) if $\Theta(fg) = \Theta(f)$ for all $g \in G$.

Let $D_G \colon G \to F$ denote the *Weyl discriminant*:

$$D_G(g) := \prod_{\alpha \in R(\mathbf{G}, \mathbf{T})} (\alpha(g) - 1). \tag{4.4.1.1}$$

Definition 4.4.4. A semisimple element $g \in G$ is called *regular* if $D_G(g) \neq 0$. Let G_{rs} denote the set of regular semisimple elements of G.

Let (π, V) be an admissible representation of G. Then, for any $f \in C_c^\infty(G)$, the operator

$$\pi(f) := \int_G f(g)\pi(g)\mathrm{d}g \tag{4.4.1.2}$$

is an operator of finite rank acting on V.

The mapping

$$\Theta_\pi \colon f \mapsto \mathrm{tr}(\pi(f)) \quad \text{for } f \in C_c^\infty(G) \tag{4.4.1.3}$$

is a G-invariant distribution, called the (distribution) *character* of π.

Under the assumption that $\mathrm{char}(F) = 0$, Harish-Chandra showed in [49] that there is a locally integrable function Θ_π, defined for all regular semisimple elements g in G such that

$$\Theta_\pi(f) = \int_{G_{\mathrm{rs}}} f(g)\Theta_\pi(g)\mathrm{d}g, \quad \text{for all } f \in C_c^\infty(G). \tag{4.4.1.4}$$

Since G_{rs} is an open dense subset of G, we may consider Θ_π as a function on G by setting $\Theta_\pi(g) = 0$ for all elements g in the complement of the regular G_{rs}, a set of measure zero.

If π and π' are equivalent admissible representations of G, then $\Theta_\pi = \Theta_{\pi'}$.

Theorem 4.4.5. If $\pi_1, \ldots, \pi_r$ are pairwise inequivalent irreducible admissible representations of G, then the distributions $\Theta_{\pi_1}, \ldots, \Theta_{\pi_r}$ are linearly independent.

Proof. See [81, Lemma 1.13.1] or [5, Section 1.4] for a simpler proof. □

Corollary 4.4.6. Let (π, V) and (π', V') be admissible representations of finite length. Then π and π' have the same Jordan–Hölder factors if and only if $\Theta_\pi = \Theta_{\pi'}$.

Remark 4.4.7. In particular, two irreducible admissible representations π, π' are equivalent if and only if $\Theta_\pi = \Theta_{\pi'}$ (see [23, Corollary 2.20]).

If f is a function on G, then we define $\check{f}$ on G by $\check{f}(g) := f(g^{-1})$.

4.4.2 Parabolic Induction and Restriction

Let G be the F-rational points of a connected reductive algebraic group $\mathbf{G}$ defined over F. Let $\mathfrak{R}(G)$ be the category of all smooth complex representations of G. This is an abelian category admitting arbitrary coproducts. The category $\mathfrak{R}(G)$ is equivalent to the category of nondegenerate modules over the Hecke algebra $\mathcal{H}(G)$ of G. However, neither the center of G, nor the center of $\mathcal{H}(G)$, is sufficient to describe the center of the category $\mathfrak{R}(G)$.

A choice of Haar measure on G gives an identification of $\mathcal{C}_c^\infty(G)$ with the subspace of locally constant, compactly supported distributions $D \in \mathcal{C}_c^\infty(G)^*$.

Let $\mathbf{P}$ be an F-parabolic subgroup of $\mathbf{G}$. The *modulus character* δ_P of P is defined by the formula

$$\int_P f(p^{-1}p'p)m_P(p') = \delta_P(p) \int_P f(p')m_P(p'), \quad \text{for any } p \in P, \tag{4.4.2.1}$$

where m_P is a Haar measure on P. If χ is any character of a torus T, we may extend it in a trivial way to a character χ_B of B by setting $\chi(tu) := \chi(t)$ for any $t \in T$ and $u \in U$. The normality of U implies that the extended character is well-defined.

We will induce very special representations of P, indeed those representations that are trivial on U hence uniquely determined by their restriction to L. The inflation of a representation σ of L to P is the unique representation of P which restricts to σ on L and trivial on U; we denote it as $\mathrm{infl}_L^P(\sigma)$. The parabolic induction functor is the following composition

$$\mathrm{i}_{L,P}^G : \mathfrak{R}(L) \xrightarrow{\mathrm{inf}_L^P} \mathfrak{R}(P) \xrightarrow{\mathrm{Ind}_P^G} \mathfrak{R}(G). \tag{4.4.2.2}$$

The smooth induction functor Ind_P^G has a left adjoint, the restriction from G to P. To obtain the left adjoint functor to parabolic induction, we have to compose restriction with the U-coinvariants functor, which is left adjoint to inflation. Indeed, since U is normal in P, for any smooth representation (τ, V_τ) of P, the space $V(U)$ is stable under P and the quotient V_U provides a smooth representation of P that is trivial on U, hence a smooth representation of L. The parabolic restriction functor, or Jacquet (restriction) functor, is the following composition

$$r_{L,P}^G : \mathfrak{R}(G) \xrightarrow{\mathrm{Res}_L^P} \mathfrak{R}(P) \xrightarrow{U\text{-coinvariants}} \mathfrak{R}(L). \tag{4.4.2.3}$$

Proposition 4.4.8. Let $P = LU$ be a parabolic subgroup of G. Then the functors $\mathrm{i}_{L,P}^G$ and $r_{L,P}^G$ are both exact, and $r_{L,P}^G$ is the left adjoint of $\mathrm{i}_{L,P}^G$.

Theorem 4.4.9. [Bernstein] The right adjoint of $\mathrm{i}_{L,P}^G$ is the functor $r_{L,\overline{P}}^G$, where $\overline{P}$ denotes the opposite parabolic subgroup of P.

Definition 4.4.10. The set of all irreducible representations of G that are subquotients of $\mathrm{i}_{T,B}^G(\chi)$, with B a Borel subgroup of G and χ a character of a torus $T \subset B$ is called the *principal series* of representations of G.

Definition 4.4.11. A smooth representation π of G is *supercuspidal* if $r_{L,P}^G(\pi) = 0$, for any proper parabolic subgroup P of G.

Proposition 4.4.12. Let π be a smooth representation of G. The following conditions are equivalent

(1) the representation π is supercuspidal;
(2) the representation π is not a subquotient of any proper parabolically induced representation.
(3) every matrix coefficient of π has compact modulo center support.

Theorem 4.4.13. (Harish-Chandra) Any smooth irreducible π of G occurs as an irreducible component of a parabolically induced representation $\mathrm{i}_{L,P}^G(\sigma)$, where P is a parabolic subgroup of G with Levi factor L and σ is supercuspidal irreducible smmoth repfresentation of L. The G-conjugacy class $(L,\sigma)_G$ of (L,σ) is uniquely determined and is called the supercuspidal support of π, it is denoted by $\mathrm{Sc}(\pi)$.

4.4.3 Iwahori-spherical Representations

If $\mathbf{G}$ is F-split, let I be an Iwahori subgroup of G. We set

$$e_I := \frac{1}{\mathrm{meas}(I)}\delta_I \in \mathcal{H}(G). \tag{4.4.3.1}$$

Recall the convolution defined in (4.2.4.1).

Definition 4.4.14. The *Iwahori–Hecke algebra* of G, denoted by $\mathcal{H}(G \mathbin{/\!/} I)$ is the the convolution algebra consisting of all compactly supported left and right I-invariant functions on G. Then, e_I is an idempotent and $\mathcal{H}(G \mathbin{/\!/} I) = e_I * \mathcal{H}(G) * e_I$.

The algebra $\mathcal{H}(G \mathbin{/\!/} I)$ is an affine Hecke algebra.

Let (π, V) be a smooth representation of G. We denote by V^I the I-invariant vectors in V:

$$V^I := \{v \in V \ : \ \pi(g)v = v \text{ for any } g \in I\}. \tag{4.4.3.2}$$

Let $\mathfrak{R}_I(G)$ denote the full subcategory of smooth representations of G that are generated by their I-fixed vectors (i.e., $\mathcal{H}(G)V^I = V$).

Theorem 4.4.15. [27] The functor $V \mapsto V^I$ is an equivalence between $\mathfrak{R}_I(G)$ and the category of $\mathcal{H}(G \mathbin{/\!/} I)$-modules. If V is an admissible representation of G, then V^I is a finite-dimensional $\mathcal{H}(G \mathbin{/\!/} I)$-module.

The inverse functor is $U \mapsto C_c^\infty(G/I) \otimes_{\mathcal{H}(G/\!/I)} U$, where the action on the latter is by the left regular representation of G.

Furthermore, the category of $\mathcal{H}(G \mathbin{/\!/} I)$-modules is equivalent to a product of categories of certain graded affine Hecke algebra modules [61]. It is known that these equivalences induce bijections between the unitary representations in the corresponding categories [18, 19].

Remark 4.4.16. An interesting class of unitary representations are those that have a single K-type with Iwahori fixed vectors, or in terms of Borel's equivalence of categories, the Iwahori–Hecke algebra modules whose restrictions to the finite Hecke algebra are irreducible. These representations are expected to be automorphic, for example, the Speh representations for $\mathrm{GL}_n(\mathbb{Q}_p)$ (see [87]) are such one-K-type representations.

4.4.4 Unipotent Representations

4.4.4.1 Representations of Finite Reductive Groups

Let $\mathbb{G}$ be a connected reductive algebraic group defined over a finite field $\mathbb{F}_q$. We denote by $\mathbb{G}^\vee$ a connected reductive algebraic group defined over $\mathbb{F}_q$ with root datum dual to that of $\mathbb{G}$. More precisely, the decomposition is obtained as follows: As proved in [41, Corollary 7.7], for any irreducible representation τ of $\mathbb{G}(\mathbb{F}_q)$, there exists an $\mathbb{F}_q$-rational maximal torus $\mathbb{T}$ of $\mathbb{G}$ and a character of $\mathbb{T}(\mathbb{F}_q)$ such that τ occurs in the Deligne–Lusztig (virtual) character $R_{\mathbb{T}}^{\mathbb{G}}(\theta)$, that is, such that $\langle \tau, R_{\mathbb{T}}^{\mathbb{G}}(\theta)\rangle_{\mathbb{G}(\mathbb{F}_q)} \neq 0$, where the character of τ is also denoted by τ, and $\langle\ ,\ \rangle_{\mathbb{G}(\mathbb{F}_q)}$ is the usual scalar product on the space of class functions on $\mathbb{G}(\mathbb{F}_q)$:

$$\langle f_1, f_2\rangle_{\mathbb{G}(\mathbb{F}_q)} = |\mathbb{G}(\mathbb{F}_q)|^{-1} \sum_{g\in\mathbb{G}(\mathbb{F}_q)} f_1(g)\,\overline{f_2(g)}.$$

If $\theta = 1$ (the trivial character of $\mathbb{T}(\mathbb{F}_q)$), then the representation τ is called *unipotent*.

If the pairs $(\mathbb{T},\theta)$ and $(\mathbb{T}',\theta')$ are not $\mathbb{G}(\mathbb{F}_q)$-conjugate, then $R_{\mathbb{T}}^{\mathbb{G}}(\theta)$ and $R_{\mathbb{T}'}^{\mathbb{G}}(\theta')$ are orthogonal to each other with respect to $\langle , \rangle_{\mathbb{G}(\mathbb{F}_q)}$, but they may have a common constituent as they are virtual characters. This has motivated the introduction of the following weaker notion of conjugacy: The pairs $(\mathbb{T},\theta)$ and $(\mathbb{T}',\theta')$ are called geometrically conjugate if there exists $g \in \mathbb{G}$ such that $\mathbb{T}' = g\mathbb{T}g^{-1}$ and such that for any non-negative integer m we have $\theta' \circ \mathrm{N}_{\mathrm{Fr}^m|\mathrm{Fr}} = \theta \circ \mathrm{N}_{\mathrm{Fr}^m|\mathrm{Fr}} \circ \mathrm{ad}(g)$, where $\mathrm{Fr}\colon \mathbb{G} \to \mathbb{G}$ is the geometric Frobenius endomorphism associated with the $\mathbb{F}_q$-structure of $\mathbb{G}$ (hence we have $\mathbb{G}^{\mathrm{Fr}} = \mathbb{G}(\mathbb{F}_q)$), and $\mathrm{N}_{\mathrm{Fr}^m|\mathrm{Fr}}\colon \mathbb{T}^{\mathrm{Fr}^m} \to \mathbb{T}^{\mathrm{Fr}}$ is the norm map, given by $\mathrm{N}_{\mathrm{Fr}^m|\mathrm{Fr}}(t) := t \cdot \mathrm{Fr}(t) \cdot \mathrm{Fr}^2(t) \cdots \mathrm{Fr}^{m-1}(t)$.

The $\mathbb{G}(\mathbb{F}_q)$-conjugacy classes of pairs $(\mathbb{T},\theta)$ as above are in one-to-one correspondence with the $\mathbb{G}^\vee(\mathbb{F}_q)$-conjugacy classes of pairs $(\mathbb{T}^\vee, s)$ where s is a semisimple element of $\mathbb{G}^\vee(\mathbb{F}_q)$ and $\mathbb{T}^\vee$ is an $\mathbb{F}_q$-rational maximal torus of $\mathbb{G}^\vee$ containing s. Then, the Lusztig series $\mathscr{E}(\mathbb{G}, s)$ is defined to be the set of τ be an irreducible representation of $\mathbb{G}(\mathbb{F}_q)$ such that τ occurs in $R_{\mathbb{T}}^{\mathbb{G}}(\theta)$, where $(\mathbb{T},\theta)_{\mathbb{G}(\mathbb{F}_q)}$ corresponds to $(\mathbb{T}^\vee, s)_{\mathbb{G}^\vee(\mathbb{F}_q)}$.

By definition, $\mathscr{E}(\mathbb{G}, 1)$ consists only of unipotent representations.

Then the set of equivalence classes $\mathrm{Irr}(\mathbb{G}(\mathbb{F}_q)$ of irreducible representations of $\mathbb{G}(\mathbb{F}_q)$ decomposes into a disjoint union

$$\mathrm{Irr}(\mathbb{G}(\mathbb{F}_q)) = \bigsqcup_{(s)} \mathscr{E}(\mathbb{G}(\mathbb{F}_q), s), \tag{4.4.4.1}$$

where (s) is the $\mathbb{G}^\vee(\mathbb{F}_q)$-conjugacy class of a semisimple element s of $\mathbb{G}^\vee(\mathbb{F}_q)$.

Let $\mathbb{G}_s^\vee := \mathrm{Cent}_{\mathbb{G}^\vee}(s)$ denote the centralizer of s in $\mathbb{G}^\vee$. In the case when $\mathbb{G}$ has a connected center, the group $\mathbb{G}_s^\vee$ is connected. For an arbitrary $\mathbb{G}$, the group $\mathbb{G}_s^\vee$ may be disconnected, and we will start by extending the notion of Deligne–Lusztig characters to this case as follows. We denote by $\mathbb{G}_s^{\vee,\circ}$ the identity component of $\mathbb{G}_s$, and, for $\mathbb{T}^\vee$ an $\mathbb{F}_q$-rational maximal torus of $\mathbb{G}_s^{\vee,\circ}$ and $\theta^\vee$ a character of $\mathbb{T}^\vee(\mathbb{F}_q)$, we define

$$R_{\mathbb{T}^\vee}^{\mathbb{G}_s^\vee}(\theta^\vee) := \mathrm{Ind}_{\mathbb{G}_s^{\vee,\circ}(\mathbb{F}_q)}^{\mathbb{G}_s^\vee(\mathbb{F}_q)}(\theta^\vee). \tag{4.4.4.2}$$

Let $\mathscr{E}(\mathbb{G}_s^\vee(\mathbb{F}_q),1)$ be the set of irreducible components of $R_{\mathbb{T}^\vee}^{\mathbb{G}_s^\vee}(1)$.

Theorem 4.4.17. [Lusztig] There is a bijection

$$\begin{aligned}\mathscr{E}(\mathbb{G}(\mathbb{F}_q),s) &\xrightarrow{\sim} \mathscr{E}(\mathbb{G}_s^\vee(\mathbb{F}_q),1)\\ \tau &\mapsto \tau_{\mathrm{unip}},\end{aligned} \tag{4.4.4.3}$$

that sends any Deligne–Lusztig character $R_{\mathbb{T}}^{\mathbb{G}}(\theta)$ to (up to a sign) a Deligne–Lusztig character $R_{\mathbb{T}^\vee}^{\mathbb{G}_s^\vee}(1)$, where 1 denotes the trivial character of $\mathbb{T}^\vee$.

Proof. In the case when the center of $\mathbb{G}$ is connected, see [57, Theorem 4.23]. For $\mathbb{G}$ arbitrary, see [58, Section 12]. □

Moreover, the bijection (4.4.4.3) preserves cuspidality in the following sense: Let $\tau \in \mathscr{E}(\mathbb{G}(\mathbb{F}_q),s)$ be cuspidal. Then the following properties are satisfied:

1. if $s_0 \in \mathbb{G}^\vee(\mathbb{F}_q)$ is $\mathbb{G}^\vee(\mathbb{F}_q)$-conjugate of s, then the largest $\mathbb{F}_q$-split torus in the center of $\mathbb{G}_{s_0}^\vee$ coincides with the largest $\mathbb{F}_q$-split torus in the center of $\mathbb{G}^\vee$ (see [57, (8.4.5)]),
2. the unipotent representation τ_{unip} is cuspidal.

4.4.5 Classification of Unipotent Representations

Definition 4.4.18. An irreducible admissible representation V of G is *unipotent* if there is a parahoric subgroup $G_{x,0}$ of G with pro-unipotent radical $G_{x,0+}$ such that the $G_{x,0+}$-invariants in V contain a cuspidal unipotent representation of the finite reductive group $\mathbb{G}_{x,0}(\mathbb{F}_q)$.

Lusztig has established a parametrization of unipotent representations in the case where $\mathbf{G}$ is simple of adjoint type in [59, 60]. It has been extended to any G in [46, 82, 83]. This goes as follows. Let $G^\vee$ be the complex Langlands dual of G. Then the unipotent representations of G are in bijective correspondence with $G^\vee$-conjugacy classes of triples (s,u,ρ), where $s \in G^\vee$ is semisimple,

$u \in G^\vee$ is unipotent such that $sus^{-1} = u^q$, and ρ is the isomorphism class of an irreducible representation of the component group of the mutual centralizer in $G^\vee$ of s and u, such that ϱ is trivial on the center of $G^\vee$. Let $E_{s,u,\varrho}$ denote the irreducible G-representation corresponding to the indicated triple. Kazhdan and Lusztig had earlier proved [53] that the corresponding parahoric subgroup is minimal (an Iwahori subgroup) if and only if ϱ appears in the homology of the mutual fixed points of s and u on the flag manifold of $G^\vee$.

4.4.6 Depth-zero Representations

Definition 4.4.19. An irreducible smooth representation (π, V) has *depth zero* if $V^{G_{x,0+}} \neq \{0\}$ for some parahoric subgroup $G_{x,0}$ of G.

Remark 4.4.20. Every irreducible Iwahori-spherical representation has depth zero.

We first recall the following fundamental result about depth zero supercuspidal representations, which is due to Morris [65, Section 1–2], and independently Moy and Prasad [68, Proposition 6.6].

Theorem 4.4.21. Let π be an irreducible depth-zero supercuspidal representation of G. Then there exists a maximal parahoric subgroup $G_{x,0}$ such that π is compactly induced from a representation ρ of the normalizer of $G_{x,0}$ in G such that the restriction of ρ to $G_{x,0+}$ is 1-isotypic.

Example 4.4.22. Let τ be an irreducible cuspidal representation of $\mathrm{GL}_N(\mathbb{F}_q)$, let $\widetilde{\tau}$ be the inflation of τ to the maximal compact subgroup $K_0 = \mathrm{GL}_N(\mathfrak{o}_F)$ of $\mathrm{GL}_N(F)$ that was considered in (4.3.2.1). We have $K_0 = G_{x,0}$, where x is a vertex in $\mathscr{B}(\mathrm{GL}_N, F)$. The normalizer of K_0 in G is the group $F^\times K_0$. Let $\boldsymbol{\tau}$ denote the extension to $F^\times K_0$ of τ. The representation $\mathrm{c{-}Ind}_{F^\times K_0}^{\mathrm{GL}_N(F)} \boldsymbol{\tau}$ is an irreducible depth-zero supercuspidal representation of $\mathrm{GL}_N(F)$.

4.4.7 Tempered Representations

Definition 4.4.23. Let (π, V_π) be a representation of G that is unitary with respect to an inner product $\langle\ ,\ \rangle$. Given vectors v_1 and v_2 in V_π, the corresponding *matrix coefficient* of π is the function on G given by $g \mapsto \langle \pi(g)v_1, v_2\rangle$.

Definition 4.4.24. An irreducible unitary representation with square-integrable matrix coefficients is called a *square-integrable representation.* The *discrete series* of representations is the collection of square-integrable (mod the center of G) representations.

The discrete series of G splits into two classes:

(1) supercuspidal representations: irreducible unitary representations whose matrix coefficients are compactly supported (mod the center of G),
(2) generalized special representations: irreducible unitary representations whose matrix coefficients are square-integrable (mod the center of G), and which are subrepresentations of representations induced from proper parabolic subgroups of G.

Example 4.4.25. The center of $G = \mathrm{SL}_2(F)$ is $\{\pm 1\}$. It follows that discrete series consists of those irreducible unitary representations whose matrix coefficients lie in $L^2(G)$. The discrete series for G was first discussed in [47] and analyzed in considerably more detail in [80].

Definition 4.4.26. Let $P = LU$ be a parabolic subgroup of G, (π, V) an admissible representation of G, χ a smooth character of A_L (the F-points of $\mathbf{A_L}$, the maximal F-split torus lying in the center of $\mathbf{L}$), K_L an open compact subgroup of L. We define $(\mathrm{r}^G_{L,P}V)^{K_L}_\chi$ to be the set of $v \in (\mathrm{r}^G_{L,P}V)^{K_L}$ such that there exists $d \in \mathbb{Z}_{>0}$ with

$$\left(\mathrm{r}^G_{L,P}(\pi)(t) - \chi(t)\mathrm{Id}_V\right)^d \cdot v = 0 \quad \text{for all } t \in A.$$

We have

$$(\mathrm{r}^G_{L,P}V)^{K_L} = \bigoplus_\chi (\mathrm{r}^G_{L,P}V)^{K_L}_\chi,$$

where χ describes the set of smooth characters of A.

If $K'_L \subset K_L$ is another open compact subgroup of G, we have $(\mathrm{r}^G_{L,P}V)^{K_L}_\chi \subset (\mathrm{r}^G_{L,P}V)^{K'_L}_\chi$, and we set

$$(\mathrm{r}^G_{L,P}V)_\chi := \bigcup_{K_L} (\mathrm{r}^G_{L,P}V)^{K_L}_\chi,$$

where K_L runs over the set of open compact subgroups of G.

Definition 4.4.27. If $(\mathrm{r}^G_{L,P}V)_\chi \neq \{0\}$, we say that χ is an *exponent* of π for P. The set of these exponents will be denoted by $\mathrm{Exp}(A_L, \mathrm{r}^G_{L,P}V)$.

Fix a minimal parabolic subgroup $P_0 = L_0U_0 \subset P$ such that $A_L \subset A_0 := A_{L_0}$. Let $X^*(A_L)$ be the group of rational characters of A_L, and $X_*(A_L) := \mathrm{Hom}_\mathbb{Z}(X_*(A_L), \mathbb{Z})$. We write

$$\mathfrak{a}_L := X_*(A_L) \otimes_\mathbb{Z} \mathbb{R} \quad \text{and} \quad \mathfrak{a}^*_L := X^*(A_L) \otimes_\mathbb{Z} \mathbb{R}.$$

Let $\Sigma(A_0)$ be the set of roots of A_0 in $\mathfrak{g}$, and $\Sigma_{\text{red}}(A_0)$ the subset of reduced roots: $\Sigma_{\text{red}}(A_0) := \{\alpha \in \Sigma(A_0) : \alpha/m \notin \Sigma(A_0) \text{ if } m \geq 2\}$. Let Δ_0 be the set of simple roots in $\Sigma_{\text{red}}(A_0)$, and set

$$\Delta(P) := \{\alpha_{|\mathfrak{a}_L} : \alpha \in \Delta_0 \backslash \Delta_0^L\},$$

where Δ_0^L is the analog of Δ_0 where G is replaced by L.

Lemma 4.4.28. [Casselman's criterion] Let (π, V) be an irreducible smooth representation of $\mathbf{G}$ with unitary central character. Then π is square-integrable modulo the center if and only if, for any parabolic subgroup $P = LU$ of G and any $\chi \in \text{Exp}(A, \text{r}_{L,P}^G V)$, we have $\text{Re}(\chi) \in {}^+[\mathfrak{a}_L^*]_P^G$, where

$$^+[\mathfrak{a}_L^*]_P^G := \left\{ \sum_{\alpha \in \Delta(P)} m_\alpha \alpha \, : \, m_\alpha > 0 \right\}.$$

Proof. See [72, VII.1.2]. □

Proposition 4.4.29. Suppose that (π, V) is irreducible smooth and square-integrable modulo the center. Then (π, V) is unitarizable and there exists a unique $\text{fdeg}(\pi) > 0$ such that

$$\int_{G/Z_G} \langle u, \pi(g^{-1})u \rangle \cdot \langle v, \pi(g)v \rangle \text{d}\bar{g} = \text{fdeg}(\pi)^{-1} \langle \tilde{u}, v \rangle \cdot \langle \tilde{v}, u \rangle.$$

Proof. See [72, IV.3.3]. □

Definition 4.4.30. The constant $\text{fdeg}(\pi)$ is called the *formal degree* of π. (Note that $\text{fdeg}(\pi)$ depends on a choice of Haar measure on G/Z_G.)

Definition 4.4.31. If π is such that $\pi \otimes \chi$ is square integrable modulo the center for some quasi-character (i.e., one-dimensional smooth representation) χ of G, then π is said to be *essentially square-integrable modulo the center*, and we set

$$\text{fdeg}(\pi) := \text{fdeg}(\pi \otimes \chi).$$

We write

$$^+\overline{[\mathfrak{a}_L^*]}_P^G := \left\{ \sum_{\alpha \in \Delta(P)} m_\alpha \alpha \, : \, m_\alpha \geq 0 \right\}.$$

Definition 4.4.32. A representation (π, V) of G is *tempered* if it is admissible, and if for any parabolic subgroup $P = LU$ of G, every $\chi \in \text{Exp}(A, \text{r}_{L,P}^G V)$ satisfies $\text{Re}(\chi) \in {}^+\overline{[\mathfrak{a}_L^*]}_P^G$.

We denote by $\text{Irr}(G)$ the set of equivalence classes of smooth irreducible complex representations of G, by $\text{Irr}_\text{t}(G)$ the set of equivalence classes of irreducible tempered representations of G, and by $\text{Irr}_2(G)$ the set of equivalence classes of irreducible square-integrable modulo the center representations of G. Any $\pi \in \mathrm{Irr}_2(G)$ is tempered (see [72, VII.2.1]). Thus, we have

$$\mathrm{Irr}_2(G) \subset \mathrm{Irr}_{\mathrm{t}}(G) \subset \mathrm{Irr}(G). \tag{4.4.7.1}$$

Theorem 4.4.33. [Harish-Chandra] An irreducible representation π of G is tempered if and only if it occurs as an irreducible component of a parabolically induced representation $\mathrm{i}_{M,Q}^{G}(\delta)$, where Q is a parabolic subgroup of G with Levi factor M and $\delta \in \mathrm{Irr}_2(M)$. The G-conjugacy class $(M,\delta)_G$ of (M,δ) is uniquely determined and is called the *discrete support* of π.

Proof. See [72, VII.2.6]. □

Corollary 4.4.34. Any irreducible tempered representation is unitary.

Proof. It follows from Theorem 4.4.33, since any $\pi \in \mathrm{Irr}_2(G)$ is unitary (see [72, IV.3.2]), and since parabolic induction preserves unitarity. □

We write

$$\begin{aligned} \Omega(G) &:= \{(L,\sigma)_G \ : \ L \text{ Levi subgroup of } G,\, \sigma \text{ irred. supercuspidal}\} \\ \Omega^{\mathrm{t}}(G) &:= \{(M,\delta)_G \ : \ M \text{ Levi subgroup of } G,\, \delta \text{ discrete series}\}, \end{aligned} \tag{4.4.7.2}$$

and denote the corresponding *supercuspidal* and *discrete support maps* by

$$\nu\colon \mathrm{Irr}_{\mathrm{t}}(G) \to \Omega(G) \quad \text{and} \quad \nu^{\mathrm{t}}\colon \mathrm{Irr}_{\mathrm{t}}(G) \to \Omega^{\mathrm{t}}(G). \tag{4.4.7.3}$$

Theorem 4.4.35. (Harish-Chandra) Any $\pi \in \mathrm{Irr}_{\mathrm{t}}(G)$ occurs as an irreducible component of a parabolically induced representation $\mathrm{i}_{M,P}^{G}(\delta)$, where P is a parabolic subgroup of G with Levi factor M and $\delta \in \mathrm{Irr}_2(M)$.

The G-conjugacy class $(M,\delta)_G$ of (M,δ) is uniquely determined and is called the *discrete support* of π.

4.4.8 The Plancherel Formula

Fix a maximal special parahoric subgroup K_0 of G, which is in a good relative position with L_0. A simple description of K_0 can be formulated using affine buildings as follows. Let $\mathscr{B}(G)$ denote the extended affine buildings of G. The group K_0 is a parahoric subgroup associated with any special point x of $\mathscr{B}(G)$ that lies in $\mathscr{B}(M_0)$. If H is a closed subgroup of G and d_H a Haar measure on H (left or right), then we will always suppose that it is normalized, so the measure of $K_0 \cap H$ is one.

We recall the definition [48] of the Harish-Chandra γ and c-factors. For $P = MN$, a parabolic subgroup of G, we denote by $\overline{P}$ (resp $\overline{N}$) the opposite parabolic subgroup (resp. the unipotent radical of $\overline{P}$). Let $\mathrm{d}\overline{n}$ denote the Haar measure on $\overline{N}$. We extend the modulus character δ_P of P defined in (4.4.2.1) to a function δ'_P on all G, using the decomposition $G = PK_0$, by setting $\delta'_P(pk) := \delta_P(p)$ for $p \in P$ and $k \in K_0$. Then the Harish-Chandra γ-factor $\gamma(G|M)$ is defined as

$$\gamma(G|M) := \int_{\overline{N}} \delta'_{\overline{P}}(\overline{n})\mathrm{d}\overline{n} \tag{4.4.8.1}$$

We write $R(P, A_L)$ for the set of reduced roots of P with respect to A_L. Let α be a reduced root of A_L in G, let A_α denote the connected component of the kernel of α in A_L, and L_α be the centralizer of A_α in G. Then L_α is a Levi subgroup in G containing L and $K_0 \cap M_\alpha$ is a maximal compact subgroup of L_α. We set

$$c(G|M) := \gamma(G|M)^{-1} \cdot \prod_\alpha \gamma(M_\alpha|M), \tag{4.4.8.2}$$

where the product runs over all the reduced roots of A_0 in P.

The *Plancherel formula* for a p-adic reductive group G is an explicit decomposition of the trace

$$\mathcal{H}(G) \to \mathbb{C},\ f \mapsto f(1) \tag{4.4.8.3}$$

in terms of the traces of irreducible representations of G. It is an essential tool of invariant harmonic analysis on real or p-adic reductive groups.

Let Ξ be the usual spherical function on G:

$$\Xi(g) := \int_{K_0} \delta_{P_0}(gk)^{-1} dk. \tag{4.4.8.4}$$

A function $f \in \mathcal{C}(G)$ is said to be rapidly decreasing if

$$\sup_{x \in G} |f(x)|\Xi(x)^{-1}(1 + \|x\|)^N < \infty \quad \text{for all } N > 0, \tag{4.4.8.5}$$

where $\|\ \|$ is the distance on G. Let K be a compact open subgroup of G. We denote by $\mathcal{S}(G /\!/ K)$ the algebra of all complex-valued functions on G, which are rapidly decreasing and K-bi-invariant, with product given by convolution.

Definition 4.4.36. The *Schwartz algebra* $\mathcal{S}(G)$ of G is the inductive limit of the algebras $\mathcal{S}(G /\!/ K)$.

The topology on $\mathcal{S}(G /\!/ K)$ is defined by the semi-norms that appear in its definition, and the topology on $\mathcal{S}(G)$ is the inductive limit topology.

The *Plancherel theorem* describes $\mathscr{S}(G)$ in terms of its irreducible representations. Harish-Chandra has formulated a version of the Plancherel theorem that is similar to that for real groups (see [48, 49]), although he published only a sketch of the proof. The full proof was written by Waldspurger in [89]. The theorem states that the Schwartz space $\mathscr{S}(G)$ of a p-adic reductive group G is the orthogonal direct sum of subspaces formed from representations induced unitarily from discrete series of the Levi factors of parabolic subgroups. Moreover, if two such series of induced representations yield the same subspace of $\mathscr{S}(G)$, then the parabolic subgroups from which they are induced are associate, and the representations of the Levi factors are conjugate.

Let $\delta \in \mathrm{Irr}_2(M)$ and $\mathcal{O}^{\mathrm{t}}$ as in (4.7.1.1).

- Let $\mathscr{P}$ be the set of pairs $(Q = MN, \mathcal{O}^{\mathrm{t}})$, where Q is a semi-standard parabolic subgroup of G and $\mathcal{O}$ as earlier.
- Two pairs $(Q_1 = M_1N_1, \mathcal{O}^{\mathrm{t}}_1)$ and $(Q_2 = M_2N_2, \mathcal{O}^{\mathrm{t}}_2)$ in $\mathscr{P}$ are associated if there exists $w \in W^G$ such that $w \cdot M_1 = M_2$ and $w\mathcal{O}^{\mathrm{t}}_1 = \mathcal{O}^{\mathrm{t}}_2$.
- We fix a set $\tilde{\mathscr{P}}$ of representatives in $\mathscr{P}$ for the classes of association.
- We write $W(M, \mathcal{O}^{\mathrm{t}}) := \left\{n \in \mathrm{N}_G(M) \ : \ w(\mathcal{O}^{\mathrm{t}}) = \mathcal{O}^{\mathrm{t}}\right\} / M$.
- For each $f \in C_c^\infty(G)$, the function $f^\vee$ is defined by

$$f^\vee(g) := f(g^{-1}) \quad \text{for } g \in G.$$

- Recall: if (π, V) is a smooth representation of G, given $f \in C_c^\infty(G)$, the operator $\pi(f) \colon V \to V$ is defined by

$$\pi(f)v := \int_G f(g)\pi(g)v\mathrm{d}g, \quad \text{for } v \in V.$$

Theorem 4.4.37. [89] For each $f \in C_c^\infty(G)$ and each $g \in G$, we have

$$f(g) = \sum_{(Q,\mathcal{O}^{\mathrm{t}}) \in \tilde{\mathscr{P}}} c_{G,M} |W(M, \mathcal{O}^{\mathrm{t}})|^{-1} \int_{\mathcal{O}^{\mathrm{t}}} \mu_{G|M}(\delta) \cdot \mathrm{fdeg}(\delta) \cdot \theta_\delta^G(\lambda(g) f^\vee)\, \mathrm{d}\delta,$$

where

$$\theta_\delta^G := \mathrm{trace}(\pi(f)), \quad \text{where } \pi = i_{M,Q}^G(\delta),$$

$c_{G,M}$ is a constant depending on G and M, and $\mu_{G|M}$ is the Plancherel measure defined by Harish-Chandra.

The explicit Plancherel measure related to the components in the Schwartz space decomposition has been found in several cases (see [14, 56, 77–79]).

4.5 Decomposition of the Smooth Dual

4.5.1 Support Maps

We recall that

$$\Omega^{\mathrm{t}}(G) := \{(M,\delta)_G \; : \; M \text{ Levi subgroup of } G \text{ and } \delta \text{ discrete series of } M\}, \quad (4.5.1.1)$$

$$\Omega(G) := \{(L,\rho)_G \; : \; L \text{ Levi subgroup of } G \text{ and } \rho \text{ irred. supercuspidal of } L\}, \quad (4.5.1.2)$$

and denote by

$$\nu^{\mathrm{t}} \colon \mathrm{Irr}_{\mathrm{t}}(G) \to \Omega^{\mathrm{t}}(G) \quad \text{and} \quad \nu \colon \mathrm{Irr}_{\mathrm{t}}(G) \to \Omega(G) \quad (4.5.1.3)$$

the corresponding discrete support and supercuspidal support maps.

Proposition 4.5.1. [Compatibility of the discrete and supercuspidal support maps] The following diagram commutes:

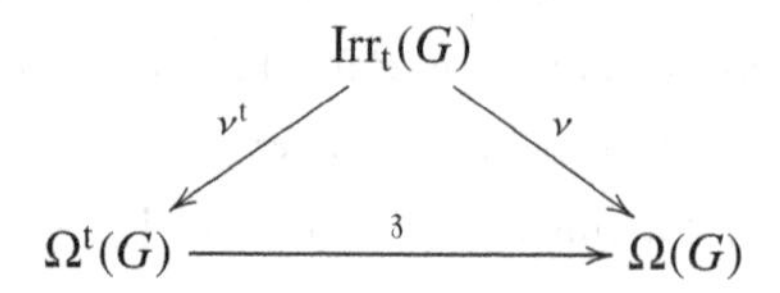

where $\mathfrak{z}\colon (M,\omega)_G \mapsto$ supercuspidal support of ω.

Remark 4.5.2. For $G = \mathrm{GL}_n(F)$, the map $\mathfrak{z}$ is injective and ν^{t} is bijective (since unitary induction is irreducible for $\mathrm{GL}_n(F)$).

4.5.2 The Bernstein Center

In the study of a real, reductive Lie group, the center of the universal enveloping algebra of the Lie algebra plays an important role. If $\mathfrak{g}$ is the Lie algebra of a real Lie group $\mathscr{G}$, then the category of $\mathfrak{g}$-modules is equivalent to the category of $\mathfrak{U}(\mathfrak{g})$-modules, where $\mathfrak{U}(\mathfrak{g})$ is the enveloping algebra of $\mathfrak{g}$. The center of this category, that is, the algebra of all natural transformations of the identity functor, is isomorphic to the center of $\mathfrak{U}(\mathfrak{g})$. In particular, the center of $\mathfrak{g}$ is insufficient for describing the center of the category.

A similar situation occurs in the p-adic case:

Definition 4.5.3. [21] The *Bernstein center* $\mathfrak{Z}(G)$ of G is the algebra of endomorphisms of the identity functor of the category $\mathfrak{R}(G)$.

There are two realizations of $\mathfrak{Z}(G)$. The most explicit description is in terms of algebraic geometry: $\mathfrak{Z}(G)$ is the algebra of regular functions on $\Omega(G)$.

A second description of $\mathfrak{Z}(G)$ is in terms of distributions: One can view $\mathfrak{Z}(G)$ as the convolution algebra of G-invariant distributions, which are *essentially compact*: A distribution D is called *essentially compact* if for all functions f in the space $C_c^\infty(G)$, we have that the functions $f * D$ and $D * f$ are compactly supported. An elementary example of an essentially compact distribution is the delta distribution associated to a central element of G. If G is semisimple, the characters of irreducible supercuspidal representations of G are other examples of elements of $\mathfrak{Z}(G)$.

The algebra $\mathfrak{Z}(G)$ is an analogue of $\mathfrak{Z}(\mathfrak{U}(\mathfrak{g}))$ for a reductive p-adic group G. Let $\mathscr{U}(G)$ denote the algebra of essentially compact distributions on G. Like $\mathfrak{U}(\mathfrak{g})$, the algebra $\mathscr{U}(G)$, has a natural adjoint operation $*$. Moreover, the category $\mathfrak{R}(G)$ is also equivalent to the category of nondegenerate $\mathscr{U}(G)$-modules.

Let $\mathscr{H}(G)'$ denote the algebra of distributions D such that for all $f \in C_c^\infty(G)$, we have that $D * f \in C_c^\infty(G)$. We have

$$\mathscr{H}(G) \subset \mathscr{U}(G) \subset \mathscr{H}(G)' \quad \text{and} \quad \mathfrak{Z}(\mathscr{U}(G)) = \mathfrak{Z}(\mathscr{H}(G)') = \mathfrak{Z}(G).$$

The algebra $\mathscr{H}(G)'$ does not have an adjoint operation.

Remark 4.5.4. There is a natural map of $\mathscr{H}(G)$ into the algebra $\mathrm{End}_{\mathbb{C}}(C_c^\infty(G))$, but it is not an isomorphism (see [69, Section 3.2]).

Let L be a Levi subgroup of a parabolic subgroup P of G.

Definition 4.5.5. A character $\chi\colon L \to \mathbb{C}^\times$ is *unramified* if χ is trivial on every compact subgroup of L.

Let $\mathfrak{X}_{\mathrm{nr}}(L)$ denote the group of unramified characters of L, and $\mathfrak{X}_{\mathrm{unr}}(L)$ the subgroup of the unitary unramified characters. The group $\Lambda(L) := L/L^1$ is a free abelian of rank r equal to the rank of a maximal split torus in the center of **L**. A character of L is unramified if it is trivial on L^1. Clearly, $\mathfrak{X}_{\mathrm{nr}}(L) \simeq \mathrm{Hom}_{\mathbb{Z}}(\Lambda(L), \mathbb{C}^\times)$ (see [72, Section V.2.4] for more details). In particular, $\mathfrak{X}_{\mathrm{nr}}(L)$ has the structure of a complex algebraic group, which is isomorphic to $(\mathbb{C}^\times)^r$.

Let σ be an irreducible supercuspidal smooth representation of L and $\mathscr{O}$ the set of equivalence classes of representations L of the form $\sigma \otimes \chi$, with $\chi \in \mathfrak{X}_{\mathrm{nr}}(L)$. We write $\mathfrak{s} := (L, \mathscr{O})_G$ for the G-conjugacy class of the pair $(L, \mathscr{O})$ and $\mathfrak{B}(G)$ for the set of such classes $\mathfrak{s}$. We set $\mathfrak{s}_L := (L, \mathscr{O})_L$.

We denote by $\mathfrak{R}^{\mathfrak{s}}(G)$ the full subcategory of $\mathfrak{R}(G)$ whose objects are the representations (π, V) such that every irreducible G-subquotient of π is equivalent to a subquotient of a parabolically induced representation $\mathrm{i}_{L,P}^G(\sigma')$, where $\mathrm{i}_{L,P}^G$ is the functor of normalized parabolic induction and $\sigma' \in \mathscr{O}$. The categories

$\mathfrak{R}^{\mathfrak{s}}(G)$ are indecomposable and split the full smooth category $\mathfrak{R}(G)$ in a direct product (see [21, Proposition 2.10]):

$$\mathfrak{R}(G) = \prod_{\mathfrak{s}\in\mathfrak{B}(G)} \mathfrak{R}^{\mathfrak{s}}(G). \tag{4.5.2.1}$$

Let $\mathrm{Irr}^{\mathfrak{s}}(G)$ denote the set of irreducible objects of the categoty $\mathfrak{R}^{\mathfrak{s}}(G)$. As a direct consequence of (4.5.2.1), we have

$$\mathrm{Irr}(G) = \prod_{\mathfrak{s}\in\mathfrak{B}(G)} \mathrm{Irr}^{\mathfrak{s}}(G). \tag{4.5.2.2}$$

Let $\mathfrak{s} = [L,\sigma]_G \in \mathfrak{B}(G)$. We denote by L^1 the intersection of kernels of unramified characters of L. Let (σ_1, V_1) be an irreducible component of the restriction of σ to L^1. We denote by $\mathrm{c{-}Ind}_{M^1}^{M}$ the functor of compact induction. As noticed in [73, Sections 1.2], the isomorphism class of

$$\Pi^{\mathfrak{s}_L} := \mathrm{c{-}Ind}_{L^1}^{L}(\sigma_1, V_1) \tag{4.5.2.3}$$

is independent of the choice of (σ_1, V_1). It was shown by Bernstein that

$$\Pi^{\mathfrak{s}} := \mathrm{i}_{L,P}^{G}(\Pi^{\mathfrak{s}_L}) \tag{4.5.2.4}$$

is a progenerator of $\mathfrak{R}^{\mathfrak{s}}(G)$ (see [73, 1.6]). Hence, by [73, Section 1.1], the functor $V \mapsto \mathrm{Hom}_G(\Pi^{\mathfrak{s}}, V)$ is an equivalence from $\mathfrak{R}^{\mathfrak{s}}(G)$ to the category of right modules of the algebra $\mathcal{H}^{\mathfrak{s}}(G) := \mathrm{End}_G(\Pi^{\mathfrak{s}})$:

$$\mathfrak{R}^{\mathfrak{s}}(G) \approx \mathcal{H}^{\mathfrak{s}}(G) - \mathrm{Mod}. \tag{4.5.2.5}$$

4.5.3 Theory of Types

We fix a Haar measure on G, write $\mathcal{H}(G)$ for the space of locally constant, compactly supported functions $f\colon G \to \mathbb{C}$ and view $\mathcal{H}(G)$ as a $\mathbb{C}$-algebra via convolution relative to the Haar measure. Let (ρ, V_ρ) be a smooth representation of a compact open subgroup J of G. We define $\mathcal{H}(G,\rho)$ to be the space of compactly supported functions $f\colon G \to \mathrm{End}_G(V_\rho)$ such that

$$f(jgj') = \rho(j)f(g)\tilde{\rho}(j'), \quad \text{where } j, j' \in J \text{ and } g \in G. \tag{4.5.3.1}$$

The convolution product gives $\mathcal{H}(G,\rho)$ the structure of a unitary associative $\mathbb{C}$-algebra.

Let $e_\rho \in \mathcal{H}(G)$ be the function defined by

$$e_\rho(g) := \begin{cases} \frac{\dim\rho}{\mathrm{meas}(J)}\mathrm{trace}(\rho(g)) & \text{if } g \in J, \\ 0 & \text{if } g \in G,\ g \notin J. \end{cases}$$

Then e_ρ is idempotent, and $e_\rho \star \mathscr{H}(G) \star e_\rho$ is a sub-algebra of $\mathscr{H}(G)$ with unit e_ρ.

Bushnell and Kutzko defined in [33, (2.12)] a canonical isomorphism:

$$\mathscr{H}(G,\rho) \otimes_{\mathbb{C}} \mathrm{End}_{\mathbb{C}}(V_\rho) \to e_\rho \star \mathscr{H}(G) \star e_\rho.$$

The algebras $\mathscr{H}(G,\rho)$ and $e_\rho \star \mathscr{H}(G) \star e_\rho$ are therefore canonically Morita equivalent. Hence, we get an equivalence of categories of right modules:

$$\mathscr{H}(G,\rho) - \mathrm{Mod} \approx e_\rho \star \mathscr{H}(G) \star e_\rho - \mathrm{Mod}. \tag{4.5.3.2}$$

We write $\mathfrak{R}_\rho(G)$ for the full sub-category of $\mathfrak{R}(G)$ whose objects are those V satisfying $V = \mathscr{H}(G) \star e_\rho \star V$, that is, $\mathfrak{R}_\rho(G)$ is generated over G by the subspace $e_\rho \star V$.

Definition 4.5.6. [33, (4.2)] Let $\mathfrak{S}$ be a finite subset of $\mathfrak{B}(G)$. A pair (J,ρ) is an $\mathfrak{S}$-*type* for G if the following property is satisfied

$$\rho \text{ occurs in the restriction of } \pi \in \mathrm{Irr}(G) \text{ if and only if } \pi \in \mathfrak{S}. \tag{4.5.3.3}$$

If $\mathfrak{S}$ has only one element $\mathfrak{s}$, we then refer to an $\mathfrak{s}$-type, rather than an $\{\mathfrak{s}\}$-type.

Example 4.5.7. The pair $(I,1)$, where I is an Iwahori subgroup of G and 1, the trivial representation of I, is an $\mathfrak{s}$-type for $\mathfrak{s} = [T,1]_G$, where T is the group of F-points of a split torus and 1, the trivial representation of T.

More generally, let x be a point in the Bruhat–Tits building of G, let τ be a cuspidal irreducible representation of $G_x/G_{x,0+}$, and let $\widetilde{\tau}$ denote the inflation of τ to G_x.

Theorem 4.5.8. [66, Theorem 4.9] The pair $(G_x,\widetilde{\tau})$ is an $\mathfrak{s}$-type for G, where $\mathfrak{s} = [L,\sigma]_G$ such that σ has depth-zero.

Moy and Prasad introduced in [67] the notion of *depth* of a smooth representation, which is an important invariant of such representations (see for instance [7, 15]). We recall it later, following [40] in which the reliance on $\mathfrak{g}$ has been removed from the definition. Let (π,V) be a smooth representation of G. By [40, Lemma 5.2.1], there exists $\mathrm{d}(\pi) \in \mathbb{Q}$ with the following properties.

(1) If $(x,r) \in \mathscr{B}(\mathbf{G},F) \times \mathbb{R}_{\geq 0}$ such that $V^{G_{x,r+}}$ is nontrivial, then $r \geq \mathrm{d}(\pi)$.
(2) There exists a $x' \in \mathscr{B}(\mathbf{G},F)$ such that $V^{G_{x',\mathrm{d}(\pi)+}}$ is nontrivial.

Then the *depth of* π is defined to be the least nonnegative real number $\mathrm{d}(\pi)$ for which there exists an $x \in \mathscr{B}(\mathbf{G},F)$ such that $V^{G_{x,\mathrm{d}(\pi)+}}$ is nontrivial.

Beginning with the work of Howe, there are two constructions of supercuspidal representations by the process of compact induction from open, compact (modulo center) subgroups. One is the construction of Bushnell and Kutzko for $\mathrm{GL}_n(F)$ in [31], which has been extended to the setting of classical groups by Stevens in [86] and to inner forms of $\mathrm{GL}_n(F)$ by Sécherre and Stevens in [76]. It has been shown to be exhaustive for $\mathrm{GL}_n(F)$ and its inner forms and for p odd in the case of classical groups.

The second is the construction in [90]. Generalizing the previous work of Adler, J.-K. Yu used the Moy–Prasad filtration to construct positive-depth supercuspidal representations of G, in the case where $\mathbf{G}$ splits over a tamely ramified extension. When the residual characteristic p of F does not divide the order of the Weyl group of $\mathbf{G}$, Yu's construction gives all supercuspidal irreducible representations of G (see [43, 54]). However, it was recently noticed by Spice that the proofs of two essential results in this construction, [90, Proposition 14.1 and Theorem 14.2], are not correct, due to the usage of a misstated lemma in a reference. Although [44, Theorem 3.1] shows that Yu's construction still produces irreducible supercuspidal representations, it appeared to be important to restore the validity of Yu's results. This was recently done in [45, Th. 4.1.13] by modifying Yu's construction with a certain quadratic character. By [63], for essentially tame irreducible supercuspidal representations, Yu's construction in [90] coincides with Bushnell–Kutzko's construction in [31].

Using the theory of G-covers introduced by Bushnell and Kutzko in [33], Kim and Yu showed in [55] that Yu's construction of supercuspidal representations can also be used to obtain types by omitting some of the conditions that Yu imposed on his input data. Every smooth irreducible representation of G contains such a type if p does not divide the order of the Weyl group (see [43]). In [16], we have similarly extended the "twisted Yu's construction" of [45] to arbitrary irreducible smooth representations.

Let $\mathfrak{S}$ be a finite subset of $\mathfrak{B}(G)$. By [33, (4.3)], if (J,ρ) is an $\mathfrak{S}$-type for G, then

$$\mathfrak{R}_\rho(G) = \prod_{\mathfrak{s}\in\mathfrak{S}} \mathfrak{R}^{\mathfrak{s}}(G), \tag{4.5.3.4}$$

as a product of subcategories. In particular, if (J,ρ) is an $\mathfrak{s}$-type for G, then we have $\mathfrak{R}_\rho(G) = \mathfrak{R}^{\mathfrak{s}}(G)$, and, by [33, Theorem 3.5, 3.6, 3.12], the latter is equivalent to the category of modules of $\mathcal{H}(G,\rho)$:

$$\mathfrak{R}^{\mathfrak{s}}(G) \approx \mathcal{H}(G,\rho) - \mathrm{Mod}. \tag{4.5.3.5}$$

By combining (4.5.3.5) and (4.5.3.6), we obtain an equivalence of right modules

$$\mathcal{H}^{\mathfrak{s}}(G) - \mathrm{Mod} \approx \mathcal{H}(G,\rho) - \mathrm{Mod}. \tag{4.5.3.6}$$

In many cases, the equivalence (4.5.3.6) can be promoted into an algebra isomorphism between $\mathrm{End}_G(\Pi^{\mathfrak{s}}_G)$ and $\mathscr{H}(G,\rho)$, as we will show in Corollary 4.5.12.

Property 4.5.9. Let $\mathfrak{s}_L = [L,\sigma]_L \in \mathfrak{B}(L)$, where σ is of the form $\sigma = \mathrm{c\!-\!Ind}_{\widetilde{J}_L}^{L} \widetilde{\rho}_L$ for a representation $\widetilde{\rho}_L$ of an open, compact mod center subgroup $\widetilde{J}_L$ of L, such that $\widetilde{J}_L$ is contained in the normalizer $\mathrm{N}_L(J_L)$ of J_L in L, and

1. the representation $\rho_L := \mathrm{Res}^{\widetilde{J}_L}_{J_L} \widetilde{\rho}_L$ is irreducible, where J_L is the unique maximal compact subgroup of $\widetilde{J}_L$;
2. any element of L that intertwines the representation ρ_L lies in $\widetilde{J}_L$, that is,

$$\left\{\ell \in L \,:\, \mathrm{Hom}_{J_L^\ell \cap J_L}(\mathrm{Res}^{J_L}_{J_L \cap \widetilde{J}_L^\ell} \rho_L^\ell, \mathrm{Res}^{J_L}_{J_L \cap \widetilde{J}_L^\ell} \rho_L) \neq \{0\}\right\} \subset \widetilde{J}_L. \quad (4.5.3.7)$$

Any $\mathfrak{s}_L \in \mathfrak{B}(L)$, for any Levi subgroup L of G, satisfies Property 4.5.9 if:

- G is $\mathrm{GL}_N(F)$, by [31, (8.4.1), (6.1.2)],
- G is $\mathrm{SL}_N(F)$ by [32] and [33, p.604 (a)],
- p is odd and G is a classical group, by [64],

By [33, (5.4)], if $\mathfrak{s}_L$ satisfies Property 4.5.9, then the pair (J_L, ρ_L) is an $\mathfrak{s}_L$-type.

Proposition 4.5.10. If $\mathfrak{s}_L = [L,\sigma]_L \in \mathfrak{B}(L)$ satisfies Property 4.5.9, then the pro-generators $\Pi^{\mathfrak{s}_L}$ and $\mathrm{c\!-\!Ind}_{J_L}^{L} \rho_L$ are isomorphic.

Proof. Since ρ_L is an $\mathfrak{s}_L$-type, by (4.5.3.3), the representation ρ_L occurs in the restriction of σ to J_L. Thus, by Frobenius reciprocity, σ is a quotient of $\mathrm{c\!-\!Ind}_{J_L}^{L} \rho_L$. Again by Frobenius reciprocity, $\mathrm{c\!-\!Ind}_{J_L}^{L^1} \rho_L$ contains at least one irreducible summand of $\mathrm{Res}^L_{L^1}\sigma$, say σ_1. We will show that $\mathrm{c\!-\!Ind}_{J_L}^{L^1} \rho_L$ is irreducible. Since it is semisimple, it is enough to show that $\dim \mathrm{Hom}_{L^1}(\mathrm{c\!-\!Ind}_{J_L}^{L^1} \rho_L, \mathrm{Res}^L_{L^1}\sigma) = 1$. By Frobenius reciprocity, we have

$$\begin{aligned}\mathrm{Hom}_{L^1}(\mathrm{c\!-\!Ind}_{J_L}^{L^1} \rho_L, \mathrm{Res}^L_{L^1}\sigma) &= \mathrm{Hom}_L(\mathrm{c\!-\!Ind}_{J_L}^{L} \rho_L, \sigma)\\ &= \mathrm{Hom}_L(\mathrm{c\!-\!Ind}_{J_L}^{L} \rho_L, \mathrm{c\!-\!Ind}_{\widetilde{J}_L}^{L} \widetilde{\rho}_L) \quad (4.5.3.8)\\ &= \mathrm{Hom}_{J_L}(\rho_L, \mathrm{Res}^L_{J_L} \circ \mathrm{c\!-\!Ind}_{\widetilde{J}_L}^{L} \widetilde{\rho}_L).\end{aligned}$$

By the Mackey formula, we have

$$\mathrm{Res}^L_{J_L} \circ \mathrm{c\!-\!Ind}_{\widetilde{J}_L}^{L} \widetilde{\rho}_L = \bigoplus_{\ell} \mathrm{c\!-\!Ind}^{J_L}_{J_L \cap \widetilde{J}_L^\ell} \circ \mathrm{Res}^{\widetilde{J}_L}_{J_L \cap \widetilde{J}_L^\ell} \widetilde{\rho}_L^\ell,$$

where the sum is taken over a set of double coset representatives ℓ in $J_L\backslash L/\widetilde{J}_L$., where $\widetilde{J}_L^\ell := \ell^{-1}\widetilde{J}_L\ell$. For a fixed ℓ, by Frobenius reciprocity, we have

$$\begin{aligned}&\mathrm{Hom}_{J_L}(\rho_L,\mathrm{c-Ind}_{J_L\cap\widetilde{J}_L^\ell}^{J_L}\circ\mathrm{Res}_{J_L\cap\widetilde{J}_L^\ell}^{\widetilde{J}_L}\widetilde{\rho}_L^\ell)\\&\simeq\mathrm{Hom}_{J_L\cap\widetilde{J}_L^\ell}(\mathrm{Res}_{J_L\cap\widetilde{J}_L^\ell}^{J_L}\rho_L,\mathrm{Res}_{J_L\cap\widetilde{J}_L^\ell}^{\widetilde{J}_L}\widetilde{\rho}_L^\ell).\end{aligned}$$

We set ${}^\ell J_L = \ell\widetilde{J}_L\ell^{-1}$. Since the group ${}^\ell J_L\cap\widetilde{J}_L$ is compact, it is contained in the maximal compact subgroup of $\widetilde{J}_L$, that is, J_L. It follows that ${}^\ell J_L\cap\widetilde{J}_L = {}^\ell J_L\cap J_L$. Therefore,

$$\begin{aligned}&\mathrm{Hom}_{J_L}(\rho_L,\mathrm{c-Ind}_{J_L\cap\widetilde{J}_L^\ell}^{J_L}\circ\mathrm{Res}_{J_L\cap\widetilde{J}_L^\ell}^{\widetilde{J}_L}\widetilde{\rho}_L^\ell)\\&\simeq\mathrm{Hom}_{J_L\cap J_L^\ell}(\mathrm{Res}_{J_L\cap J_L^\ell}^{J_L}\rho_L,\mathrm{Res}_{J_L\cap J_L^\ell}^{J_L}\rho_L^\ell).\end{aligned}$$

By (4.5.3.7), if the space $\mathrm{Hom}_{J_L\cap J_L^\ell}(\mathrm{Res}_{J_L\cap J_L^\ell}^{J_L}\rho_L,\mathrm{Res}_{J_L\cap J_L^\ell}^{J_L}\rho_L^\ell)$ is nonzero, then we have $\ell\in\widetilde{J}_L$. Since $\widetilde{J}_L\subset \mathrm{N}_L(J_L)$, this space, if nonzero, is equal to $\mathrm{Hom}_{J_L}(\rho_L,\rho_L^\ell)$, which implies that ℓ is a representative of the trivial coset in $J_L\backslash L/\widetilde{J}_L$. By (4.5.3.8), we get

$$\begin{aligned}\mathrm{Hom}_L(\mathrm{c-Ind}_{J_L}^L\rho_L,\sigma)&=\mathrm{Hom}_{L^1}(\mathrm{c-Ind}_{J_L}^{L^1}\rho_L,\mathrm{Res}_{L^1}^L\sigma)\\&=\mathrm{Hom}_{J_L}(\rho_L,\rho_L)=\mathbb{C}.\end{aligned}$$

Thus, $\mathrm{c-Ind}_{J_L}^{L^1}\rho_L$ is irreducible, and hence is equal to σ_1. By (4.5.2.3), we have

$$\Pi^{\mathfrak{s}_L}=\mathrm{c-Ind}_{L^1}^L\sigma_1=\mathrm{c-Ind}_{L^1}^L(\mathrm{c-Ind}_{J_L}^{L^1}\rho_L)=\mathrm{c-Ind}_{J_L}^L\rho_L.$$

□

4.5.4 Covers and Applications

Let $P = LU$ be a parabolic subgroup of G with Levi factor L, and let $\overline{P} = L\overline{U}$ be the opposite parabolic subgroup. A compact open subgroup J of G is said to *decompose with respect* to $(U,L,\overline{U})$ if $J = (J\cap U)\cdot(J\cap L)\cdot(J\cap\overline{U})$. Let J (resp. J_L) be a compact open subgroup of G (resp. L), and ρ (resp. ρ_L) be an irreducible smooth representation of J (resp. J_L). The pair (J,ρ) is called a *G-cover* of the pair (J_L,ρ_L) (see [25, 33]) if for any opposite pair of parabolic subgroups $P = LU$ and $\overline{P} = L\overline{U}$ with with Levi factor L, we have

1. J decomposes with respect to $(U,L,\overline{U})$;
2. $\rho|_{J_L} = \rho_L$ and $J\cap U, J\cap\overline{U}\subset\ker(\rho)$;
3. for any smooth representation V of G, the natural map from V to its Jacquet module V_U induces an injection on V^ρ the (J,ρ)-isotypic subspace of V.

By [33, Theorem 8.3], if (J_L, ρ_L) is an $\mathfrak{s}_L$-type for L, then any G-cover of (J_L, ρ_L) is an $\mathfrak{s}$-type for G. If V_J is a smooth representation of J, we denote by $(V_J)^\rho$ the ρ-isotypic part of V_J, that is, the sum of all J-invariant subspaces of V_J that are isomorphic to ρ. Let $\mathrm{r}^G_{J,\rho} : \mathfrak{R}(G) \to \mathfrak{R}(J)$ be the functor defined by

$$\mathrm{r}^G_{J,\rho}(V) := (\mathrm{Res}^G_J(V))^\rho. \tag{4.5.4.1}$$

The left adjoint of $\mathrm{r}^G_{J,\rho}$ is the functor $\mathrm{i}^G_{J,\rho} : \mathfrak{R}(J) \to \mathfrak{R}(G)$ defined by

$$\mathrm{i}^G_{J,\rho}(V) := (\mathrm{c\!-\!Ind}^G_J(V))^\rho. \tag{4.5.4.2}$$

By [33, Theorem 7.9(iii)], we have

$$\mathrm{Res}^J_{J_L} \circ \mathrm{r}^G_{J,\rho} = \mathrm{r}^L_{J_L,\rho_L} \circ \mathrm{r}^G_{L,P}. \tag{4.5.4.3}$$

By taking the left adjoint of (4.5.4.4), we obtain

$$\mathrm{i}^G_{J,\rho} \circ \mathrm{Ind}^J_{J_L} = \mathrm{i}^G_{L,\overline{P}} \circ \mathrm{i}^L_{J_L,\rho_L}. \tag{4.5.4.4}$$

Theorem 4.5.11. Let (J_L, ρ_L) be a $\mathfrak{s}_L$-type for $\mathfrak{s}_L \in \mathfrak{B}(L)$, which has the property that $\Pi^{\mathfrak{s}_L}_L \simeq \mathrm{c\!-\!Ind}^L_{J_L} \rho_L$, and let (J, ρ) be a G-cover of (J_L, ρ_L). Then we have

$$\Pi^{\mathfrak{s}}_G \simeq \mathrm{c\!-\!Ind}^G_J \rho, \tag{4.5.4.5}$$

and, as a consequence,

$$\mathrm{End}_G(\Pi^{\mathfrak{s}}_G) \simeq \mathscr{H}(G, \rho). \tag{4.5.4.6}$$

Proof. By Frobenius reciprocity, and since $\mathrm{Res}^J_{J_L} \rho = \rho_L$, we have

$$\mathrm{Hom}_J(\mathrm{Ind}^J_{J_L} \rho_L, \rho) = \mathrm{Hom}_{J_L}(\rho_L, \mathrm{Res}^J_{J_L} \rho) = \mathrm{Hom}_{J_L}(\rho_L, \rho_L) = \mathbb{C}.$$

Thus, $(\mathrm{c\!-\!Ind}^L_{J_L}(\rho_L))^\rho = \rho$. It gives

$$\mathrm{i}^G_{J,\rho} \circ \mathrm{Ind}^J_{J_L}(\rho_L) = \left(\mathrm{c\!-\!Ind}^G_J \left(\mathrm{Ind}^J_{J_L}(\rho_L)\right)^\rho\right) = \mathrm{c\!-\!Ind}^G_J \rho.$$

On the other hand, since $\Pi^{\mathfrak{s}_L}_L \simeq \mathrm{c\!-\!Ind}^M_{J_L}(\rho_L, V_{\rho_L})$, we have

$$\mathrm{i}^G_{L,\overline{P}} \circ \mathrm{i}^L_{J_L,\rho_L}(\rho_L) = \mathrm{i}^G_{L,\overline{P}}(\mathrm{c\!-\!Ind}^L_{J_L} \rho_L) \simeq \mathrm{i}^G_{L,\overline{P}}(\Pi^{\mathfrak{s}_L}_L) = \Pi_{\mathfrak{s}}.$$

Then (4.5.4.5) follows from (4.5.4.4). □

Corollary 4.5.12. If (J_L, ρ_L) is an $\mathfrak{s}_L$-type for $\mathfrak{s}_L = [L, \sigma]_L \in \mathfrak{B}(L)$ such that Property 4.5.9 is satisfied, then $\mathrm{End}_G(\Pi^{\mathfrak{s}}_G) \simeq \mathscr{H}(G, \rho)$.

Proof. Follows from the combination of Proposition 4.5.10 and Theorem 4.5.11. □

4.6 On the ABPS Conjecture

4.6.1 Twisted Extended Quotients

Let Γ be a group acting on a topological space X and let Γ_x denote the stabilizer in Γ of $x \in X$. Let $\natural$ be a collection of 2-cocycles

$$\natural_x : \Gamma_x \times \Gamma_x \to \mathbb{C}^\times, \tag{4.6.1.1}$$

such that $\natural_{\gamma x}$ and $\gamma_* \natural_x$ define the same class in $H^2(\Gamma_{\gamma x}, \mathbb{C}^\times)$, where $\gamma_* : \Gamma_x \to \Gamma_{\gamma x}$ sends γ' to $\gamma\gamma'\gamma^{-1}$. Let $\mathbb{C}[\Gamma_x, \natural_x]$ be the group algebra of Γ_x *twisted by* $\natural_x$, which is defined to be the $\mathbb{C}$-vector space $\mathbb{C}[\Gamma_x, \natural_x]$ with basis $\{t_\gamma \ : \ \gamma \in \Gamma_x\}$ and multiplication rules $t_\gamma t_{\gamma'} := \natural_x(\gamma, \gamma') t_{\gamma\gamma'}$, for any $\gamma, \gamma' \in \Gamma_x$. We set

$$\widetilde{X}_\natural := \{(x, \tau) \ : \ x \in X, \, \tau \in \operatorname{Irr} \mathbb{C}[\Gamma_x, \natural_x]\}, \tag{4.6.1.2}$$

and topologize $\widetilde{X}_\natural$ by decreeing that a subset of $\widetilde{X}_\natural$ open if its projection to the first coordinate is open in X.

We require, for every $(\gamma, x) \in \Gamma \times X$, a definite algebra isomorphism

$$f_{\gamma,x} : \mathbb{C}[\Gamma_x, \natural_x] \to \mathbb{C}[\Gamma_{\gamma x}, \natural_{\gamma x}] \tag{4.6.1.3}$$

satisfying the conditions

(a) if $\gamma x = x$, then $f_{\gamma,x}$ is conjugation by an element of $\mathbb{C}[\Gamma_x, \natural_x]^\times$;
(b) $f_{\gamma',\gamma x} \circ f_{\gamma,x} = f_{\gamma'\gamma,x}$ for all $\gamma', \gamma \in \Gamma$ and $x \in X$.

Define a Γ-action on $\widetilde{X}_\natural$ by $\gamma \cdot (x, \tau) := (\gamma x, \tau \circ f_{\gamma,x}^{-1})$. The *twisted extended quotient* of X by Γ with respect to $\natural$ is defined to be

$$(X//\Gamma)_\natural := \widetilde{X}_\natural / \Gamma. \tag{4.6.1.4}$$

In the case when the 2-cocycles $\natural_x$ are trivial, we write simply $X//\Gamma$ for $(X//\Gamma)_\natural$ and refer to it as the (spectral) *extended quotient* of X by Γ.

4.6.2 Formulation of the ABPS Conjecture

Let $\mathfrak{s} = [L, \sigma]_G \in \mathfrak{B}(G)$. The group

$$\mathfrak{X}_{\mathrm{nr}}(L, \sigma) := \{\chi \in \mathfrak{X}_{\mathrm{nr}}(L) \ : \ \sigma \otimes \chi \cong \sigma\} \tag{4.6.2.1}$$

is finite. Thus, there is a bijection

$$\mathfrak{X}_{\mathrm{nr}}(L)/\mathfrak{X}_{\mathrm{nr}}(L, \sigma) \to \operatorname{Irr}^{\mathfrak{s}_L}(L) \ : \ \chi \mapsto \sigma \otimes \chi, \tag{4.6.2.2}$$

which endows $\operatorname{Irr}^{\mathfrak{s}_L}(L)$ with the structure of a complex torus. Up to isomorphism, this torus depends only on $\mathfrak{s}$, and it is known as the *Bernstein torus* $T_\mathfrak{s}$ attached to $\mathfrak{s}$. We note that $T_\mathfrak{s}$ is only an algebraic variety, it is not endowed with a natural multiplication map. In fact, it does not even possess an unambiguous "unit", because in general there is no preferred choice of an element $\sigma \in \mathfrak{s}_L$.

The group $W_G(L) := \mathrm{N}_G(L)/L$ acts on $\mathrm{Irr}(L)$ by

$$w \cdot \pi = [\bar{w} \cdot \pi \, : \, l \mapsto \pi(\bar{w}^{-1} l \bar{w})] \quad \text{for any lift } \bar{w} \in \mathrm{N}_G(L) \text{ of } w \in W_G(L). \tag{4.6.2.3}$$

Bernstein also associated with $\mathfrak{s}$ the finite group

$$W_{\mathfrak{s}} := \mathrm{N}_G(\mathfrak{s}_L)/L = \{w \in W_G(L) \, : \, w \cdot \sigma = \sigma \otimes \chi \text{ for some } \chi \in \mathfrak{X}_{\mathrm{nr}}(L)\}. \tag{4.6.2.4}$$

The group $W_{\mathfrak{s}}$ acts naturally on $T_{\mathfrak{s}}$, by automorphisms of algebraic varieties. Let t be a point $t \in T_{\mathfrak{s}}$. We denote by $W_{\mathfrak{s},t}$ the stabilizer in $W_{\mathfrak{s}}$ of t.

Let $\mathrm{Irr}^{\mathfrak{s}}_{\mathfrak{t}}(G)$ be the subset of tempered representations in $\mathrm{Irr}^{\mathfrak{s}}(G)$, and let $T_{\mathfrak{s},\mathrm{c}}$ denote the maximal compact real subtorus of $T_{\mathfrak{s}}$. The action of $W_{\mathfrak{s}}$ on $T_{\mathfrak{s}}$ preserves $T_{\mathfrak{s},\mathrm{c}}$, and we can form the twisted extended quotient $(T_{\mathfrak{s},\mathrm{c}}//W_{\mathfrak{s}})_{\natural}$.

The ABPS conjecture from [6, Section 15] and [8, Conjecture 2] in its roughest form asserts that there exists a family of 2-cocycles

$$\natural_t : W_{\mathfrak{s},t} \times W_{\mathfrak{s},t} \to \mathbb{C}^\times \qquad t \in T_{\mathfrak{s}}, \tag{4.6.2.5}$$

and a bijection

$$\mathrm{Irr}^{\mathfrak{s}}(G) \longleftrightarrow (T_{\mathfrak{s}} /\!/ W_{\mathfrak{s}})_{\natural}, \tag{4.6.2.6}$$

which restricts to a bijection between $\mathrm{Irr}^{\mathfrak{s}}_{\mathfrak{t}}(G)$ and $(T_{\mathfrak{s},\mathrm{c}} /\!/ W_{\mathfrak{s}})_{\natural}$.

The set $\mathrm{Irr}_{\mathrm{cusp}}(L)$ of supercuspidal L-representations is stable under the $W_G(L)$-action (4.6.2.3). The definitions of $W_{\mathfrak{s}}$ and of extended quotients imply that for a fixed Levi subgroup L of G there is a natural bijection

$$\bigsqcup\nolimits_{\mathfrak{s}=[L,\sigma]_G} (T_{\mathfrak{s}} /\!/ W_{\mathfrak{s}})_{\natural} \to \left(\mathrm{Irr}_{\mathrm{cusp}}(L) /\!/ W_G(L)\right)_{\natural}. \tag{4.6.2.7}$$

In view of (4.5.2.1), the ABPS conjecture can also be formulated in terms of a bijection

$$\mathrm{Irr}(G) \longleftrightarrow \bigsqcup\nolimits_L \left(\mathrm{Irr}_{\mathrm{cusp}}(L) /\!/ W_G(L)\right)_{\natural}, \tag{4.6.2.8}$$

where L runs through a set of representatives for the G-conjugacy classes of Levi subgroups of G. In this version, our conjecture asserts that $\mathrm{Irr}(G)$ is determined by a much smaller set of data, namely the supercuspidal representations of Levi subgroups L of G, and the actions of the Weyl groups $W_G(L)$ on those.

Remark 4.6.1. A Galois version of the ABPS conjecture for enhanced L-parameters was formulated and proved in [12, Theorem 9.3].

4.6.3 Known Cases

The ABPS conjecture was proved for representations in the principal series of an arbitrary split p-adic groups in [11], for all smooth representations of inner forms of $\mathrm{GL}_n(F)$ in [10], for the exceptional p-adic group G_2 in [4, 17], and for pure inner forms of classical groups in [13]: In all these cases, it only involves extended quotients (i.e., no twisting is needed).

In general, it is expected that the group cohomology classes $\natural_t \in H^2(W_{\mathfrak{s},t}, \mathbb{C}^\times)$ reflect the character of $Z(G^\vee)^{W_F}$ which via the Kottwitz isomorphism determines how G is an inner twist of a quasi-split group. In particular, $\natural$ should be trivial whenever G is quasi-split. The simplest known example of a nontrivial cocycle involves a non split inner form of $\mathrm{SL}_{10}(F)$ [9, Example 5.5]. That example also shows that it is sometimes necessary to use *twisted* extended quotients in the ABPS conjecture: It is indeed the case in the proof of the ABPS Conjecture for inner forms of SL_n, see [10].

A general proof was recently obtained in [84]. Let $\mathfrak{s} = [L,\sigma]_G \in \mathfrak{B}(G)$. The action of every element w of $W_\mathfrak{s}$ can be lifted to a transformation $\tilde{w}$ of $\mathfrak{X}_{\mathrm{nr}}(L)$. Let $W(L,\sigma,\mathfrak{X}_{\mathrm{nr}}(L))$ be the group of permutations of $\mathfrak{X}_{\mathrm{nr}}(L)$ generated by $\mathfrak{X}_{\mathrm{nr}}(L,\sigma)$ and the $\tilde{w}$'s. It satisfies

$$W(L,\sigma,\mathfrak{X}_{\mathrm{nr}}(L))/\mathfrak{X}_{\mathrm{nr}}(L,\sigma) \simeq W_\mathfrak{s}.$$

Let $K(B) := \mathbb{C}(\mathfrak{X}_{\mathrm{nr}}(L))$ denote the quotient field of $B := \mathbb{C}[\mathfrak{X}_{\mathrm{nr}}(L)]$. By [84, Corollary 5.8], there is a 2-cocycle

$$\kappa\colon W(L,\sigma,\mathfrak{X}_{\mathrm{nr}}(L))^2 \to \mathbb{C}^\times \qquad (4.6.3.1)$$

and of an algebra isomorphism

$$K(B) \otimes_B \mathrm{End}_G(\mathrm{i}_{L,P}^G(V_B)) \simeq K(B) \rtimes \mathbb{C}[W(L,\sigma,\mathfrak{X}_{\mathrm{nr}}(L)),\kappa],$$

where $\mathbb{C}[W(L,\sigma,\mathfrak{X}_{\mathrm{nr}}(L)),\kappa]$ is the twisted group algebra of $W(L,\sigma,\mathfrak{X}_{\mathrm{nr}}(L))$, and the symbol $\rtimes$ denotes a crossed product: As vector space, it just means the tensor product, and the multiplication rules on that are determined by the action of $W(L,\sigma,\mathfrak{X}_{\mathrm{nr}}(L))$ on $K(B)$. The cocycle κ is trivial on $W_\mathfrak{s}^\circ$.

For any $\chi \in \mathfrak{X}_{\mathrm{nr}}(L)$, let $W_G^{\mathfrak{s},\chi\otimes\sigma}$ denote the stabilizer in $W_\mathfrak{s}$ of $\chi \otimes \sigma$ and $\natural_\chi$ the restriction to $W_G^{\mathfrak{s},\chi\otimes\sigma}$ of the 2-cocycle $\natural_u$ defined in [84, (6.18)]. Let $(T_\mathfrak{s}/\!/W_\mathfrak{s})_\natural$ be the twisted extended quotient with respect to the collection $\natural$ of the 2-cocycles $\natural_\chi$, as defined in Section 4.6.1.

Theorem 4.6.2. There is a bijection

$$\xi_G^\mathfrak{s}\colon \mathrm{Irr}^\mathfrak{s}(G) \longrightarrow (T_\mathfrak{s}/\!/W_\mathfrak{s})_\natural. \qquad (4.6.3.2)$$

Proof. We set $B := \mathbb{C}[L/L_1]$ and $V_B := V \otimes_{\mathbb{C}} B$. Then, $\mathrm{i}_{L,P}^G(V_B)$ is also a progenerator of $\mathfrak{R}^{\mathfrak{s}}(G)$. Thus, we have the equivalence of categories of right modules

$$\begin{array}{rccc} \mathscr{E}\colon & \mathfrak{R}^{\mathfrak{s}}(G) & \longrightarrow & \mathrm{End}_G(\mathrm{i}_{L,P}^G(V_B))-\mathrm{Mod} \\ & \mathcal{V} & \mapsto & \mathrm{Hom}_G(\mathrm{i}_{L,P}^G(V_B),\mathcal{V}) \end{array}. \tag{4.6.3.3}$$

By [84, Theorem 9.7], there are bijections

$$\mathrm{Irr}^{\mathfrak{s}}(G) \xrightarrow{\mathscr{E}} \mathrm{Irr}(\mathrm{End}_G(\mathrm{i}_{L,P}^G(V_B)) \xrightarrow{\zeta} \mathrm{Irr}(\mathbb{C}[\mathfrak{X}_{\mathrm{nr}}(L)] \rtimes \mathbb{C}[W(L,\sigma,\mathfrak{X}_{\mathrm{nr}}(L)),\natural_{\mathfrak{s}}], \tag{4.6.3.4}$$

where $\natural_{\mathfrak{s}}$ is a 2-cocycle of $W_{\mathfrak{s}}$, and $\mathscr{E}$ is induced by the equivalence of categories defined in (4.6.3.3). On the other hand, [84, Lemma 9.8] shows that $\mathrm{Irr}(\mathbb{C}[\mathfrak{X}_{\mathrm{nr}}(L)] \rtimes \mathbb{C}[W(L,\sigma,\mathfrak{X}_{\mathrm{nr}}(L)),\natural_{\mathfrak{s}}])$ is canonically isomorphic to $(\mathrm{Irr}^{\mathfrak{s}_L}(L)/\!/W_{\mathfrak{s}})_{\natural}$, where $\mathfrak{s}_L := [L,\sigma]_L$. □

Corollary 4.6.3. There is a bijection

$$\mathrm{Irr}(\mathscr{H}^{\mathfrak{s}}(G)) \longrightarrow (\mathrm{Irr}^{\mathfrak{s}_L}(L)/\!/W_{\mathfrak{s}})_{\natural}. \tag{4.6.3.5}$$

Proof. It follows from the combination of (4.5.3.6) and (4.6.3.2). □

Remark 4.6.4. As observed in [84, (10.12)], if the restriction of σ to L_1 is multiplicity-free, then we have

$$\Pi_G^{\mathfrak{s}} = \left(\mathrm{i}_{L,P}^G(V_B)\right)^{\mathfrak{X}_{\mathrm{nr}}(L,\sigma)} \quad \text{and} \quad \mathrm{End}_G(\mathrm{i}_{L,P}^G(V_B)) \simeq \mathscr{H}^{\mathfrak{s}}(G) \otimes_{\mathbb{C}} \mathrm{M}_{[L:L_\sigma]}(\mathbb{C}). \tag{4.6.3.6}$$

By [74, Remark 1.6.1.3], if σ is generic, then $\sigma|_{L_1}$ is multiplicity-free.

The twisted group algebra $\mathbb{C}[W_{\mathfrak{s}},\natural_{\mathfrak{s}}]$ acts "almost" on the objects of $\mathfrak{R}^{\mathfrak{s}}(G)$ by intertwining operators: that is, these intertwining operators depend rationally on $\chi \in \mathfrak{X}_{\mathrm{nr}}(L)$, and they can have poles. In this setting, [84, Section 5] provides an isomorphism of $\mathbb{C}[\mathfrak{X}_{\mathrm{nr}}(L)]^{W_{\mathfrak{s}}}$-algebras:

$$\mathbb{C}(\mathfrak{X}_{\mathrm{nr}}(L))^{W_{\mathfrak{s}}} \otimes_{\mathbb{C}[\mathfrak{X}_{\mathrm{nr}}(L)]^{W_{\mathfrak{s}}}} \mathscr{H}^{\mathfrak{s}}(G) \simeq \mathbb{C}(\mathfrak{X}_{\mathrm{nr}}(L)) \rtimes \mathbb{C}[W_{\mathfrak{s}},\natural_{\mathfrak{s}}]. \tag{4.6.3.7}$$

Although the algebras $\mathbb{C}(\mathfrak{X}_{\mathrm{nr}}(L)) \rtimes \mathbb{C}[W_{\mathfrak{s}},\natural_{\mathfrak{s}}]$ and $\mathscr{H}^{\mathfrak{s}}(G)$ are usually not Morita equivalent, it turned out that they nevertheless share many properties. In particular, by [85, Theorem B], there exists a (canonical when (4.6.3.7) is fixed) linear bijection

$$\mathrm{HH}(\mathbb{C}(\mathfrak{X}_{\mathrm{nr}}(L)) \rtimes \mathbb{C}[W_{\mathfrak{s}},\natural_{\mathfrak{s}}]) \to \mathrm{HH}(\mathscr{H}^{\mathfrak{s}}(G)). \tag{4.6.3.8}$$

The category $\mathfrak{R}_{\mathfrak{t}}^{\mathfrak{s}}(G)$ of tempered representations in $\mathfrak{R}^{\mathfrak{s}}(G) \sim \mathrm{Mod} - \mathscr{H}^{\mathfrak{s}}(G)$ is $\mathrm{Mod} - \mathscr{S}(G)^{\mathfrak{s}}$, where $\mathscr{S}(G)^{\mathfrak{s}}$ denotes the direct summand of the Schwartz algebra $\mathscr{S}(G)$ associated with $\mathfrak{s}$ in the Bernstein decomposition of $\mathfrak{R}(G)$. This subcategory is stable under tensoring with elements of $\mathfrak{X}_{\mathrm{unr}}(G)$, and one can expect strong similarities between $\mathscr{S}(G)^{\mathfrak{s}}$ and $\mathscr{C}^{\infty}(\mathfrak{X}_{\mathrm{unr}}(L)) \rtimes \mathbb{C}[W_{\mathfrak{s}}, \natural_{\mathfrak{s}}]$.

4.6.4 Bernstein Decomposition of the Reduced C^*-algebra of G

Bernstein decomposition of $C^*_{\mathrm{r}}(G)$: We have

$$C^*_{\mathrm{r}}(G) = \bigoplus_{\mathfrak{s}\in\mathfrak{B}(G)} C^*_{\mathrm{r}}(G)^{\mathfrak{s}} \tag{4.6.4.1}$$

where the spectrum of $C^*_{\mathrm{r}}(G)^{\mathfrak{s}}$ is $\mathrm{Irr}^{\mathfrak{s}}_{\mathrm{t}}(G)$.

Conjecture 4.6.5. [5] We have

$$K_j(C^*_{\mathrm{r}}(G)^{\mathfrak{s}}) \simeq K^j_{W_{\mathfrak{s}}}(T^{\mathfrak{s}}_{\mathrm{c}}) \quad \text{for } j = 0, 1, \tag{4.6.4.2}$$

where $K^j_{W_{\mathfrak{s}}}(T^{\mathfrak{s}}_{\mathrm{c}})$ is the classical topological equivariant K-theory for the extended finite Weyl group $W_{\mathfrak{s}}$ acting on the compact torus $T^{\mathfrak{s}}_{\mathrm{c}}$.

Remark 4.6.6. For the Iwahori point $\mathfrak{s} = [T, 1]_G$, Conjecture 4.6.5 has been proved in [8, Equation (4.9)]. More generally, it should follow from the combination of [82] and [83], whenever the representation σ is unipotent.

4.7 Tempered Dual and C^*-Algebras

4.7.1 Analytic R-groups

Let δ be a square-integrable irreducible representation of M, and let $\mathscr{O}^{\mathrm{t}}$ denote the orbit of δ under $\mathfrak{X}_{\mathrm{unr}}(M)$:

$$\mathscr{O}^{\mathrm{t}} := \{\delta \otimes \chi \,:\, \chi \in \mathfrak{X}_{\mathrm{unr}}(M)\} = \mathfrak{X}_{\mathrm{unr}}(M) \cdot \delta. \tag{4.7.1.1}$$

We denote by W_δ the stabilizer of δ in $\mathrm{N}_G(M)/M$:

$$W_\delta := \{n \in \mathrm{N}_G(M) \mid {}^n\delta \simeq \delta\}/M, \tag{4.7.1.2}$$

where ${}^n\delta\colon m \mapsto \delta(n^{-1}mn)$ of L. Let Δ be the set of roots for (G, M). We have $\mu^{M_\alpha} : \mathscr{O}^{\mathrm{t}} \to \mathbb{R}^+$. The set

$$\Delta'_\delta := \left\{\alpha \in \Delta \,:\, \mu^{M_\alpha}(\delta) = 0\right\} \tag{4.7.1.3}$$

is itself a root system. We denote by W°_δ the subgroup of W_δ generated by the reflections in the roots of Δ'_δ.

Definition 4.7.1. The *Knapp–Stein R-group* (so-called *analytic R-group*) $\mathscr{R}_\delta$ of δ is defined to be

$$\mathscr{R}_\delta := \left\{ w \in W_\delta \ : \ w\alpha > 0, \text{ for any } \alpha \in \Delta'_\delta \right\}. \tag{4.7.1.4}$$

We have

$$W_\delta = W_\delta^\circ \rtimes \mathscr{R}_\delta, \tag{4.7.1.5}$$

and in turn $\mathscr{R}_\delta \simeq W_\delta / W_\delta^\circ$.

Central extensions of R-groups: The theory of the R-group can be brought to full fruition once chosen a central extension

$$1 \to \mathcal{Z}_\delta \to \widetilde{\mathscr{R}}_\delta \xrightarrow{p} \mathscr{R}_\delta \to 1$$

with the property that the 2-cocycle of $\widetilde{\mathscr{R}}_\delta$ induced by η_δ is a coboundary. Such a central extension exists, and we can then choose a map $\xi_\delta \colon \widetilde{\mathscr{R}}_\delta \to \mathbb{C}^\times$ that splits η_δ, in that:

$$\eta_\delta(w_1, w_2) = \frac{\xi_\delta(w_1 w_2)}{\xi_\delta(w_1)\xi_\delta(w_2)}, \quad \text{for any } w_1, w_2 \in \widetilde{\mathscr{R}}_\delta. \tag{4.7.1.6}$$

Theorem 4.7.2. [3] The irreducible components of $\mathrm{i}_{M,Q}^G(\delta)$ are in natural bijection with the set of irreducible representations of $\widetilde{R}_\delta$ with Z_δ-central character ζ_δ, where $\zeta_\delta \colon Z_\delta \to \mathbb{C}^\star$ is given by $\zeta_\delta(z) := \xi_\delta(z)^{-1}$ for $z \in Z_\delta$.

The extension $\widetilde{\mathscr{R}}_\delta$ comes with a central idempotent $\widetilde{p}$ acting on the group algebra of $\widetilde{\mathscr{R}}_\delta$:

$$\widetilde{p} := |\mathcal{Z}_\delta|^{-1} \sum_{z \in \mathcal{Z}_\delta} \zeta_\delta(z)^{-1} z \in \mathbb{C}[\mathcal{Z}_\delta]. \tag{4.7.1.7}$$

Then, $\widetilde{p}$ is a minimal idempotent in $\mathbb{C}[Z_\delta]$ such that $\widetilde{p}(\mathbb{C}[\widetilde{\mathscr{R}}_\delta]) \simeq \mathbb{C}[\mathscr{R}_\delta, \eta_\delta]$.

4.7.2 Decomposition of the Tempered Dual

Definition 4.7.3. The G-conjugacy class $\Theta := (M, \mathcal{O}^{\mathrm{t}})_G$ of a discrete pair $(M, \mathcal{O}^{\mathrm{t}})$ is called an *inertial discrete pair* in G. We write also $\Theta = [M, \delta]_G$ for $\delta \in \mathcal{O}^{\mathrm{t}}$. We denote by $\mathfrak{B}^2(G)$ the set of inertial discrete pairs in G.

We recall the discrete support map $\nu^{\mathrm{t}} \colon \mathrm{Irr}_{\mathrm{t}}(G) \to \Omega^{\mathrm{t}}(G)$ defined in (4.5.1.3).

Definition 4.7.4. Let Q be a parabolic subgroup of G with Levi factor L, and let $\mathrm{Irr}_{\mathrm{t}}^{\Theta}(G)$ denote the set of irreducible representations of G that occur in one of the induced representations $i_{M,Q}^G(\delta \otimes \chi)$, $\chi \in \mathfrak{X}_{\mathrm{unr}}(M)$:

$$\mathrm{Irr}_{\mathrm{t}}^{\Theta}(G) := (\nu^{\mathrm{t}})^{-1}(\Theta). \tag{4.7.2.1}$$

Theorem 4.7.5. [70] We have

$$\mathrm{Irr}_{\mathrm{t}}(G) = \bigsqcup_{\Theta \in \mathfrak{B}^2(G)} \mathrm{Irr}_{\mathrm{t}}^{\Theta}(G). \qquad (4.7.2.2)$$

Our next objective is to describe the structure of the $\mathrm{Irr}_{\mathrm{t}}^{\Theta}(G)$'s up to strong Morita equivalence.

4.7.3 Decomposition of the Reduced C^*-algebra of a p-Adic Reductive Group

Choose a left-invariant Haar measure on G and form the Hilbert space $L^2(G)$. The left regular representation ρ of $L^1(G)$ is given by $(\lambda(f))(h) := f * h$, where $f \in L^1(G)$ and $h \in L^2(G)$, and $*$ denotes convolution.

Definition 4.7.6. The *reduced C^*-algebra* of G, denoted as $C_{\mathrm{r}}^*(G)$, is the C^*-algebra generated by the image of λ.

The spectrum of $C_{\mathrm{r}}^*(G)$ may be identified with the tempered dual $\mathrm{Irr}_{\mathrm{t}}(G)$ of G. We will describe a decomposition:

$$C_{\mathrm{r}}^*(G) = \bigoplus_{\Theta \in \mathfrak{B}^2(G)} C_{\mathrm{r}}^*(G;\Theta), \qquad (4.7.3.1)$$

where $C_{\mathrm{r}}^*(G;\Theta)$ is a subalgebra of $C_{\mathrm{r}}^*(G)$ with spectrum $\mathrm{Irr}_{\mathrm{t}}^{\Theta}(G)$.

Let $\delta \in \mathrm{Irr}_2(M)$. We fix a Haar measure on G and a maximal compact subgroup K of G, with measure of total mass one. We may and will suppose that K is a good special maximal compact subgroup in "good relative position" with L (in particular, $G = KQ$). We introduce the following notations:

- $E_Q(\tau)$, for $(\tau, V_\tau) \in \mathscr{O}^{\mathrm{t}}$, is the space of functions $f\colon G \to V_\tau$ that are right-invariant under some compact open subgroup of G and satisfy

$$f(xmn) = \delta_Q^{-\frac{1}{2}}(m)(\tau)(m)^{-1} f(x), \quad \text{for } x \in G, L \in M, n \in N.$$

  and $\mathrm{I}_Q(\tau)$ the representation of G on $E_Q(\tau)$ by left translation.
- E_Q^K the space of functions $f\colon K \to V_\tau$ that are right-invariant under some open subgroup of K and satisfy

$$f(kmn) = \tau(m)^{-1} f(k), \quad \text{for } k \in K, L \in M \cap K, n \in N \cap K.$$

Let $\mathscr{H}_Q$ (resp. $\mathscr{H}_Q^K$) denote the Hilbert completion of $E_Q(\tau)$ (resp. E_Q^K) with respect to the inner product

$$\langle f, f' \rangle = \int_K \langle f(k), f'(k) \rangle_{V_\tau} dk.$$

We still write $\mathrm{I}_Q(\tau)$ for the representation of G on $\mathscr{H}_Q(\tau)$.

For $\chi \in \mathfrak{X}_{\mathrm{unr}}(M)$, we consider the linear isomorphism

$$\begin{array}{rccc} F_Q^K(\chi): & \mathcal{H}_Q(\delta \otimes \chi) & \to & \mathcal{H}_Q^K \\ & f & \mapsto & f|_K \end{array}. \tag{4.7.3.2}$$

Let $\mathrm{I}_Q^K(\delta \otimes \chi)$ denote the representation of G on $\mathcal{H}_Q^K$ such that

$$\mathrm{I}_Q^K(\delta \otimes \chi)(g) = F_Q^K(\chi) \circ \mathrm{I}_Q(\delta \otimes \chi)(g) \circ F_Q^K(\chi)^{-1}.$$

For every smooth and compactly supported function $f \in \mathcal{C}_c^\infty(G)$, the operator

$$\pi_\tau(f) := \int_G f(g)\mathrm{I}_Q^K(\tau)(g)dg.$$

is compact and depends continuously on τ. Let $\mathfrak{K}(\mathcal{H}_Q^K)$ be the algebra of compact operators on $\mathcal{H}_Q^K$. We obtain a linear map

$$\begin{array}{ccc} C_c^\infty(G) & \to & \mathcal{C}(\mathcal{O}^{\mathrm{t}}, \mathfrak{K}(\mathcal{H}_Q^K)) \\ f & \mapsto & (\tau \mapsto \pi_\tau(f)). \end{array} \tag{4.7.3.3}$$

Upon completing $C_c^\infty(G)$ to $C_{\mathrm{r}}^*(G)$, there arises a C^*-morphism

$$C_{\mathrm{r}}^*(G) \to \mathcal{C}(\mathcal{O}^{\mathrm{t}}, \mathfrak{K}(\mathcal{H}_Q^K)). \tag{4.7.3.4}$$

Let $C_{\mathrm{r}}^*(G;\Theta)$ denote the image of $C_{\mathrm{r}}^*(G)$ by this morphism.

From the earlier discussion, we obtain a morphism

$$C_{\mathrm{r}}^*(G) \to \bigoplus_{\Theta \in \mathfrak{B}^2(G)} C_{\mathrm{r}}^*(G;\Theta). \tag{4.7.3.5}$$

Theorem 4.7.7 ([70])**.** The map in (4.7.3.5) is a C^*-isomorphism.

Let $\Theta = [M,\delta]_G \in \mathfrak{B}^2(G)$. We set

$$\mathrm{N}_G(\Theta) := \{n \in \mathrm{N}_G(M) \, : \, {}^n\delta \simeq \delta \otimes \chi \;\text{ for some } \chi \in \mathfrak{X}_{\mathrm{unr}}(M)\}, \tag{4.7.3.6}$$

and define the stabilizer of $\mathcal{O}^{\mathrm{t}}$ (so called the *inertial stabilize*r of δ) as:

$$W_\Theta := \mathrm{N}_G(\Theta)/M. \tag{4.7.3.7}$$

When $F = \mathbb{R}$, Wassermann gave a simple determination of $C_{\mathrm{r}}^*(G;\Theta)$ up to strong Morita equivalence. We will see that an analogous result holds when F is p-adic, under the assumption that the action of W_Θ on $\mathcal{O}^{\mathrm{t}}$ admits a good fixed point.

We have $W_\delta \subset W_\Theta$ for any $\delta \in \mathcal{O}$.

Proposition 4.7.8. When $F = \mathbb{R}$, the action of W_Θ on $\mathcal{O}^{\mathrm{t}}$ has always a fixed point, that is, there exists $\delta \in \mathcal{O}^{\mathrm{t}}$ such that $W_\delta = W_\Theta$.

Question (open in general): When F is p-adic and $\Theta = (M,\mathcal{O}^{\mathrm{t}})_G \in \mathfrak{B}^2(G)$, does the action of W_Θ on $\mathcal{O}^{\mathrm{t}}$ always admit a fixed point?

Theorem 4.7.9. Let G be a quasi-split symplectic, orthogonal, or unitary group over a p-adic field F. Then for every $\Theta = (M,\mathcal{O}^{\mathrm{t}})_G \in \mathfrak{B}^2(G)$, the action of W_Θ on $\mathcal{O}^{\mathrm{t}}$ has a fixed point.

Proof. See [2, Theorem 1.5]. □

From now on, we fix $\Theta \in \mathfrak{B}^2(G)$ and we suppose that the action of W_Θ on $\mathcal{O}^{\mathrm{t}}$ has a fixed point, say δ.

Definition 4.7.10. The fixed point δ is *good* if for every point $\tau \in \mathcal{O}^{\mathrm{t}}$, the Knapp–Stein decompositions $W_\tau = W'_\tau \rtimes \mathscr{R}_\tau$ and $W_\delta = W'_\delta \rtimes \mathscr{R}_\delta$ are compatible in the following sense:

1. we have $W'_\tau \subset W'_\delta$,
2. and the R-group $\mathscr{R}_\tau$ is isomorphic with a subgroup of $\mathscr{R}_\delta$.

For each $w \in W_\Theta$ and every $\chi \in \mathfrak{X}_{\mathrm{unr}}(M)$, there is an operator

$$\mathscr{A}(w,\delta\otimes\chi)\colon \mathscr{H}_Q^K \to \mathscr{H}_Q^K \tag{4.7.3.8}$$

that intertwines $\mathrm{I}_Q^K(\delta\otimes\chi)$ and $\mathrm{I}_Q^K(\delta\otimes(w\chi))$ and satisfies

$$\mathscr{A}(w_1w_2,\delta\otimes\chi) = \eta_\delta(w_1,w_2)\,\mathscr{A}(w_1,\delta\otimes(w_2\chi))\,\mathscr{A}(w_2,\delta\otimes\chi),$$

for any w_1, w_2 in W_Θ and $\chi \in \mathfrak{X}_{\mathrm{unr}}(M)$, where $\eta_\delta\colon W_\Theta\times W_\Theta \to \mathbb{C}^\times$ is a 2-cocycle.

Proposition 4.7.11. For every $\tau\in\mathcal{O}^{\mathrm{t}}$, the map $r\mapsto\mathscr{A}(r,\tau)$ defines a projective representation of $\mathscr{R}_\tau$ on $\mathscr{H}_Q^K$. The multiplier of this projective representation is the restriction $\eta_\tau\colon \mathscr{R}_\tau\times\mathscr{R}_\tau \to \mathbb{C}^\times$ of the cocycle η_δ.

Theorem 4.7.12. Suppose that δ is a good fixed-point for the action of W_Θ on $\mathcal{O}^{\mathrm{t}}$. Then we have the strong Morita equivalence

$$C^*_{\mathrm{r}}(G,\Theta) \underset{\text{Morita}}{\sim} \widetilde{p}\left[\,\mathscr{C}(\mathcal{O}^{\mathrm{t}}/W'_\delta)\rtimes\widetilde{R}_\delta\,\right], \tag{4.7.3.9}$$

where $\widetilde{p}$ is the central idempotent defined in (4.7.1.7).

Proof. See [2, Theorem 1.4]. □

Remark 4.7.13. How to interpret Theorem 4.7.12:

- Strong Morita equivalence preserves spectra: If A and B are strongly Morita equivalent C^*-algebras, then their spectra are homeomorphic.

- We have a strong Morita equivalence between a highly noncommutative C^*-block and (the image under $\tilde{p}$ of) an almost commutative C^*-algebra, namely the crossed product of a commutative C^*-algebra by a finite group. This allows one to infer the topology on the tempered dual, in a way that reflects reducibility of the induced representations.

References

[1] P. Abramenko and K. S. Brown. *Buildings, graduate texts in mathematics*, vol. 248, Springer, New York, 2008.

[2] A. Afgoustidis and A.-M. Aubert. *C^*-blocks and crossed products for classical p-adic groups*, IMRN **2022** (22) (2022), 17849–17908.

[3] J. Arthur. *On elliptic tempered characters*, Acta Math. **171** (1) (1993), 73–138.

[4] A.-M. Aubert, P. Baum, and R. Plymen. *Geometric structure in the principal series of the p-adic group* G_2, Represent. Theory **15** (2011), 126–169.

[5] A.-M. Aubert, P. Baum, and R. Plymen. Geometric structure in the representation theory of reductive p-adic groups II, In: *Harmonic analysis on reductive, p-adic groups*, 71–90, Contemporary Mathematics, 543, American Mathematical Society, Providence, RI, 2011.

[6] A.-M. Aubert, P. Baum, R. J. Plymen, and M. Solleveld. *Geometric structure in smooth dual and local Langlands correspondence*, Japan. J. Math. **9** (2014), 99–136.

[7] A.-M. Aubert, P. Baum, R. J. Plymen, and M. Solleveld. *Depth and the local Langlands correspondence* (Arbeitstagung Bonn, 2013), In: Progress in Math., Birkhäuser, 2016, arXiv:1311.1606.

[8] A.-M. Aubert, P. Baum, R. J. Plymen, and M. Solleveld. *Conjectures about p-adic groups and their noncommutative geometry*, in: Around Langlands Correspondences, Contemp. Math. 691 (2017), 15–51.

[9] A.-M. Aubert, P. Baum, R. J. Plymen, and M. Solleveld. *Hecke algebras for inner forms of p-adic special linear groups*, J. Inst. Math. Jussieu **16** (2) (2017), 351–419.

[10] A.-M. Aubert, P. Baum, R. J. Plymen, and M. Solleveld. *The principal series of p-adic groups with disconnected centre*, Proc. London Math. Soc. **114** (2017) 798–854.

[11] A.-M. Aubert, P. Baum, R. J. Plymen, and M. Solleveld. *Smooth duals of inner forms of* GL_n *and* SL_n, Doc. Math. **24** (2019), 373–420.

[12] A.-M. Aubert, A. Moussaoui, and M. Solleveld. *Generalizations of the Springer correspondence and cuspidal Langlands parameters*, Manuscripta Math. **157** (1–2) (2018), 121–192.

[13] A.-M. Aubert, A. Moussaoui, and M. Solleveld. *Affine Hecke algebras for classical p-adic groups*, arXiv:2211.08196.

[14] A.-M. Aubert, and R. J. Plymen. *Plancherel measure for* $\mathrm{GL}(n,F)$ *and* $\mathrm{GL}(m,D)$*: Explicit formulas and Bernstein decomposition*, J. Number Theory **112** (1) (2005), 26–66.

[15] A.-M. Aubert and R. J. Plymen. *Comparison of the depths on both sides of the local Langlands correspondence for Weil-restricted groups*, With an appendix by Jessica Fintzen, J. Number Theory **233** (2022), 24–58.

[16] A.-M. Aubert. *Bruhat-Tits buildings, representations of p-adic groups and Langlands correspondence*, J. Algebra, to appear.

[17] A.-M. Aubert and Y. Xu. *Hecke algebras for p-adic reductive groups and local Langlands correspondences for Bernstein blocks*, special volume in the memory of Jacques Tits, Adv. Math., **436** (2024), https://doi.org/10.1016/j.aim.2023.109384.

[18] D. Barbasch and A. Moy. *Reduction to real infinitesimal character in affine Hecke algebras*, J. Amer. Math. Soc. **6** (3) (1993), 611–635.

[19] D. Barbasch and A. Moy. *Unitary spherical spectrum for p-adic classical groups*, Acta Appl. Math. **44** (1–2) (1996), 3–37, Representations of Lie groups, Lie algebras and their quantum analogues.

[20] D. Barbasch, D. Ciubotaru, and P. Trapa. *Dirac cohomology for graded affine Hecke algebras*, Acta Math. **209** (2) (2012), 197–227.

[21] J. Bernstein. *Le "Centre" de Bernstein* (rédigé par P. Deligne), Representations of reductive groups over a local field, 1984, pp. 1–32.

[22] J. Bernstein. Draft of: Representations of p-adic groups, Lectures by Joseph Bernstein Harvard University, Fall 1992, Notes by Karl E. Rumelhart.

[23] J. Bernstein and A. Zelevinsky. *Representations of the group* $\mathrm{GL}(n,F)$*, where F is a local non-Archimedean field*, Uspehi Mat. Nauk. **31** (3) (1976), 5–70.

[24] J. Bernstein and A. Zelevinsky. *Induced representations of reductive p-adic groups. I*, Ann. Sci. École Norm. Sup. **10** (4) (1977), 441–472.

[25] C. Blondel, *Critère d'injectivité pour l'application de Jacquet*, C. R. Acad. Sci. Paris Sér. I Math. **325** (11) (1997), 1149–1152.

[26] A. Borel. *Linear algebraic groups*, 2nd edition. Graduate Texts in Mathematics, 126. Springer-Verlag, New York, 1991.

[27] A. Borel. *Admissible representations of a semi-simple group over a local field with vectors fixed under an Iwahori subgroup*, Invent. Math. **35** (1976), 233–259.

[28] N. Bourbaki. *Éléments de mathématique. Groupes et algèbres de Lie, Chapitres 4-5-6*, Springer, Berlin, 2007.

[29] F. Bruhat and J. Tits. *Groupes réductifs sur un corps local*, Inst. Hautes Études Sci. Publ. Math. **41** (1972), 5–251.

[30] C. J. Bushnell and G. Henniart. *The local Langlands conjecture for GL(2). Grundlehren der mathematischen Wissenschaften [Fundamental Principles of Mathematical Sciences]*, vol. 335. Springer-Verlag, Berlin, 2006.

[31] C. Bushnell and P. Kutzko. *The admissible dual of* $\mathrm{GL}(N)$ *via compact open subgroups*, vol. **129**, Annals of Math Studies, Princeton University Press, Princeton, NJ, 1993.

[32] C. Bushnell and P. Kutzko. *The admissible dual of* $\mathrm{SL}(N)$*. I*, Ann. Sci. École Norm. Sup. **26** (2) (1993), 261–280.

[33] C. Bushnell and P. Kutzko. *Smooth representations of reductive p-adic groups: structure theory via types*, Proc. London Math. Soc. **77** (3) (1998), 582–634.

[34] P. Cartier. *Representations of p-adic groups: a survey*, in: Automorphic forms, representations and L-functions (Proceedings of Symposia in Pure Mathematics, Oregon State University, Corvallis, OR, 1977), Part 1, 111–155, Proceedings

of Symposia in Pure Mathematics, XXXIII, American Mathematical Society, Providence, RI, 1979.

[35] W. Casselman. Introduction to the theory of admissible representations of p-adic reductive groups, 1995.

[36] S.S. Chern and F. Hirzebruch. eds, *Note concerning Jacques Tits*, Wolf Prize in mathematics, vol. **2**, 703–754, World Scientific Publishing Co. Inc., River Edge, NJ, 2001.

[37] D. Ciubotaru, E. Opdam, and P. Trapa. *Algebraic and analytic Dirac induction for graded affine Hecke algebras*, J. Inst. Math. Jussieu **13** (3) (2014), 447–486.

[38] D. Ciubotaru. *Spin representations of Weyl groups and the Springer correspondence*, J. ReineAngew. Math. **671** (2012) 199–222.

[39] D. Ciubotaru and P. Trapa. *Characters of Springer representations on elliptic conjugacy classes*, Duke Math. J. **162** (2) (2013), no. 2 201–223.

[40] S. DeBacker. *Some applications of Bruhat-Tits theory to harmonic analysis on a reductive p-adic group*, Michigan Math. J. **50** (2) (2002), 241–261.

[41] P. Deligne, and G. Lusztig. *Representations of reductive groups over finite fields*, Ann. Math. **103** (1) (1976), 103–161.

[42] F. Digne and J. Michel. *Representations of finite groups of Lie type, London Mathematical Society student texts*, vol. **95**, Cambridge University Press, Cambridge, 2020.

[43] J. Fintzen. *Types for tame p-adic groups*, Ann. Math. **193** (1) (2021), 303–346.

[44] J. Fintzen. *On the construction of tame supercuspidal representations*, Compos. Math. **157** (12) (2021), 2733–2746.

[45] J. Fintzen, T. Kaletha, and L. Spice. *A twisted Yu construction, Harish-Chandra characters, and endoscopy*, Duke Math. J. 172 (12) (2023), 2241–2301.

[46] Y. Feng, E. Opdam, and M. Solleveld. *Supercuspidal unipotent representations: L-packets and formal degrees*, J. Ec. Polytec. Math. **7** (2020), 1133–1193.

[47] I. M. Gelfand and M. I. Graev. *The group of matrices of second order with coefficients in a locally compact field and special functions on locally compact fields*, Uspekhi Mat. Nauk **18** (1963), 29–99.

[48] Harish-Chandra, *Harmonic analysis on reductive p-adic groups*, Proceedings of Symposia in Pure Mathematics., vol. **26**, American Mathematical Society, Providence, RI, 1973, pp. 167–192.

[49] Harish-Chandra, *Admissible invariant distributions on reductive p-adic groups*, Lie Theories and Their Applications, Queen's Papers in Pure Appl. Math., vol. **48**, Queen's University Kingston, Ontario, 1978, pp. 281–347.

[50] J. E. Humphreys, *Reflection groups and coxeter groups, Cambridge studies in advanced mathematics*, vol. **29**, Cambridge University Press, Cambridge, 1990.

[51] T. Kaletha and G. Prasad. *Bruhat–Tits theory: A new approach*, New Math. Monographs, vol. **44**, Cambridge University Press, Cambridge, 2023.

[52] T. Kaletha and O. Taibi. *The local Langlands conjecture*, Proceedings of Symposia in Pure Mathematics (to appear).

[53] D. Kazhdan and G. Lusztig. *Proof of the Deligne-Langlands conjecture for Hecke algebras*, Invent. Math. **87** (1987) 153–215.

[54] J.-L. Kim. *Supercuspidal representations: An exhaustion theorem*, J. Amer. Math. Soc. **20** (2007), 273–320.

[55] J.-L. Kim and J.-K. Yu *Construction of tame types*. Representation theory, number theory, and invariant theory, 337–357, Progress in Mathematics, 323, Birkhäuser/Springer, Cham, 2017.

[56] P. Kutzko and L. Morris. *Explicit Plancherel theorems for $\mathcal{H}(q_1, q_2)$ and* $\mathrm{SL}_2(F)$, Pure and Applied Mathematics Quarterly **5**(1) (2009), 435–467.

[57] G. Lusztig. Characters of reductive groups over a finite field, Ann. Math. Stud., Princeton, NJ, 1984.

[58] G. Lusztig. *On the representations of reductive groups with disconnected center*, Astérisque **168** (1988), 157–166.

[59] G. Lusztig. *Classification of unipotent representations of simple p-adic groups* Internat. Math. Res. Notices **1995** (11) (1995), 517–589.

[60] G. Lusztig. *Classification of unipotent representations of simple p-adic groups. II*, Represent. Theory **6** (2002), 243–289.

[61] G. Lusztig. *Affine Hecke algebras and their graded version*, **2** (3) (1989), 599–635.

[62] A. Mayeux *Bruhat-Tits theory from Berkovich's point of view. Analytic filtrations*, Ann. H. Lebesgue **5** (2022), 813–839.

[63] A. Mayeux and Y. Yamamoto. *Comparing Bushnell-Kutzko and Sécherre's constructions of types for* GL_N *and its inner forms with Yu's construction*, arXiv:2112.12367.

[64] M. Miyauchi and S. Stevens. *Semisimle types for p-adic classical groups*, Mathematische Annalen **358** (2014), 257–288.

[65] L. Morris. *Tamely ramified supercuspidal representations*, Ann. Scien. Éc. Norm. Sup. 4e série **29** (1996), 639–667.

[66] L. Morris. *Level zero G-types*, Compositio Mathematica **118** (1999), 135–157.

[67] A. Moy and G. Prasad. *Unrefined minimal K-types for p-adic groups*, Invent. Math. **116** (1994), 393–408.

[68] A. Moy and G. Prasad. *Jacquet functors and unrefined minimal K-types*, Comment. Math. Helv. **71** (1) (1996), 98–121.

[69] A. Moy and M. Tadić. Some algebras of essentially compact distributions of a reductive p-adic group. *In: Harmonic analysis, group representations, automorphic forms and invariant theory*, 247–275, Lecture Notes Series, Institute for Mathematical Sciences, National University of Singapore, vol. **12**, World Scientific Publishing, Hackensack, NJ, 2007.

[70] R. J. Plymen. *Reduced C^*-algebra for reductive p-adic groups*, J. Funct. Anal. **88** (2) (1990), 251–266.

[71] B. Rémy, A. Thuillier, and A. Werner. *An intrinsic characterization of Bruhat-Tits buildings inside analytic groups*, Michigan Math. J. **72** (2022), 543–557.

[72] D. Renard. Représentations des groupes réductifs p-adiques, Cours Spécialisés, **17**, Société Mathématique de France, Paris, 2010. pp.vi+332. ISBN: 978-285629-278-5.

[73] A. Roche. *Parabolic induction and the Bernstein decomposition*, Compositio Math. **134** (2) (2002), 113–133.

[74] A. Roche. The Bernstein decomposition and the Bernstein centre. In: *Ottawa lectures on admissible representations of reductive p-adic groups*, Fields Institute Monographs, vol. **26**, American Mathematical Society, Providence, RI, 2009, 3–52.

[75] M. Ronan. Lectures on Buildings: Updated and Revised, University of Chicago Press, Chicago, IL, 2009.

[76] V. Sécherre and S. Stevens. *Représentations lisses de* $\mathrm{GL}_m(D)$ *IV: représentations supercuspidales*, J. Inst. Math. Jussieu **7** (2008), 527–574.

[77] F. Shahidi. *Fourier transforms of intertwining operators and Plancherel measures for* $\mathrm{GL}(n)$, Amer. J. Math. **106**(1) (1984), 67–111.

[78] F. Shahidi. *A proof of Langlands' conjecture on Plancherel measures; complementary series for p-adic groups*, Ann. Math. **132**(2) (1990), 273–330.

[79] F. Shahidi. *Langlands' conjecture on Plancherel measures for p-adic groups.* In: "Harmonic analysis on reductive groups", Birkhäuser Boston, Boston, MA, 1991, pp. 277–295.

[80] J. A. Shalika. Representations of the two by two unimodular group over local fields. In: *Contributions to automorphic forms, geometry, and number theory*, Johns Hopkins University Press, Baltimore, MD, 2004, pp. 1–38.

[81] A. J. Silberger. *Introduction to harmonic analysis on reductive p-adic Groups*, Princeton University Press, Princeton, NJ, 1979.

[82] M. Solleveld. *Topological K-theory of affine Hecke algebras*, Ann. K-Theory **3** (2018), 395–460.

[83] M. Solleveld. *On unipotent representations of ramified p-adic groups*, Represent. Theory **27** (2023), 669–716.

[84] M. Solleveld. *Endomorphism algebras and Hecke algebras for reductive p-adic groups*, J. Algebra **606** (2022), 371–470.

[85] M. Solleveld. *Hochschild homology of reductive p-adic groups*, J. Noncommutative Geometry, to appear.

[86] S. Stevens. *The supercuspidal representations of p-adic classical groups*, Invent. Math. **172** (2008), 289–352.

[87] M. Tadic. *Classification of unitary representations in irreducible representations of general linear group (non-Archimedean case)*, Ann. Sci. École Norm. Sup. **19** (3) (1986), 335–382.

[88] J. Tits. *Buildings of spherical type and finite BN-pairs*, Lecture Notes in Mathematics, vol. **386**, Springer-Verlag, Berlin-New York, 1974.

[89] J.-L. Waldspurger. *La formule de Plancherel pour les groupes p-adiques (d'après Harish-Chandra)*, J. Inst. Math. Jussieu 2**2** (2003), 235–333.

[90] J.-K. Yu. *Construction of tame supercuspidal representations*, J. Amer. Math. Soc. **14** (3) (2001), 579–622.

[91] J.-K. Yu. *Smooth models associated to concave functions in Bruhat-Tits theory.* In: "Autour des schémas en groupes, III", 227–258, Panor. Synthèses, vol. **47**, Soc. Math. France, Paris, 2015.

Index